T0146802

DIRTY ELECTRICITY

ELECTRIFICATION AND THE DISEASES OF CIVILIZATION
SECOND EDITION

SAMUEL MILHAM, MD, MPH

iUniverse, Inc.
Bloomington

DIRTY
ELECTRICITY

ELECTRIFICATION AND THE DISEASES OF CIVILIZATION
SECOND EDITION

iUniverse Star
an iUniverse, Inc. imprint

iUniverse books may be ordered through booksellers or by contacting:

iUniverse
1663 Liberty Drive
Bloomington, IN 47403
www.iuniverse.com
1-800-Authors (1-800-288-4677)

ISBN: 978-1-938908-18-7 (sc)
ISBN: 978-1-938908-19-4 (e)

Library of Congress Control Number: 2012916755

Printed in the United States of America

iUniverse rev. date: 11/21/2012

*To Eddie O'Gorman, founder of the UK charity
Children with Leukemia, and to the memory of his
children, Paul and Jean, and his wife, Marion, all of
whom died of electromagnetic field-related diseases.*

"Those who ignore history are destined to repeat it."

—Edmund Burke (1729–1797)

CONTENTS

Acknowledgments   ix

Preface   xi

1 The Early Days   1

2 Albany Medical College (AMC)   4

3 Internship   8

4 Residency   14

5 New York State Department Health   20

6 University of Hawaii and the Washington State
Department of Health   27

7 On My Own   44

8 Cancer at the La Quinta Middle School   55

9 The Diseases of Civilization   69

10 Future Investigations   84

11 What To Do   95

Appendix   103

References   107

Second edition text is in bold type

FIGURES

FIGURE 1   Radial tire iron filings   40

FIGURE 2   Radial tire x-ray   41

FIGURE 3   T and B cell counts in microwave exposed and sham exposed rats   50

FIGURE 4   Pheochromocytoma hospitalizations WA   51

FIGURE 5   Childhood leukemia mortality rate by single years of age for census years 1920–1960   53

FIGURE 6   Oscilloscope tracing: channel 1 is 60-Hz AC line voltage; channel 2 is channel 1 X 10 with the 60-Hz removed with a high pass filter   58

FIGURE 7   Layout of La Quinta Middle School showing classrooms measuring above 2,000 units on the Graham-Stetzer meter   63

FIGURE 8   Percent of farm and non-farm dwellings with electric service 1920–1956 U.S. Census Bureau data   71

FIGURE 9   All causes   72

FIGURE 10   Cardiovascular   73

FIGURE 11   Malignant neoplasms   74

FIGURE 12   Diabetes   75

FIGURE 13   Urban excess mortality   76

FIGURE 14   Vista del Monte cell tower   79

FIGURE 15   The thirty-two classroom Graham Stetzer readings were averaged over quartiles ranked by distance from the cell tower   81

ACKNOWLEDGMENTS

All science is based on the cumulative ideas and work of others. Since much of my work as an epidemiologist is based on the vital registration system of statistics and population data, I'm deeply indebted to the scores of people who filled out, collected, and tabulated the millions of United States birth, death, and census records over the last century. The 1961 paper showing the emergence of the childhood leukemia age peak in the early part of the twentieth century by Michael Court-Brown and Richard Doll, as well as a graph of the time trend of United States electrification published by Jesse Ausubel and Cesare Marchetti, and finally the identification, characterization, and measurement of dirty electricity by Martin Graham and David Stetzer, were necessary precursors to my study at the La Quinta Middle School that identified dirty electricity as a potent universal carcinogen. Without the *Desert Sun* newspaper article by Mike Perrault about the La Quinta teachers' cancers, none of this new knowledge would have emerged.

At each stop on my fifty-year odyssey there were people who helped and supported me. At the New York State Health Department, Alan

Gittlesohn rescued me from an unfulfilling public health residency and worked on a number of projects with me. In Hawaii, Bob Worth was the best boss and colleague I could have asked for. At the Washington State Health Department, much of my work was facilitated and improved by Eric Ossiander. Our childhood leukemia project was critical to everything that followed. Eric has also given me a continuing bridge to Washington State data during my twenty-year retirement.

Over the years, the late William Ross Adey, MD, was always available to answer technical questions about electromagnetic fields. Louis Slesin, Ph.D., editor of *Microwave News,* has for decades been an important conduit into the latest electromagnetic field (EMF) research.

Retirement meant leaving professional networks behind. With no office, students, colleagues, library, or professional meetings, most of my research efforts have, of necessity, been a solitary enterprise for the last twenty years. It has taken every bit of the last fifty years for all of the pieces of this puzzle to fall into place to reveal the amazing picture of an invisible, hidden exposure contributing to our modern "diseases of civilization."

Thanks to George Nedeff of iUniverse for guidance and advice on publishing, and to B. Blake Leavitt for making this book more readable. Magda Havas produced most of the figures in the book. Sherry Milham offered valuable suggestions and proofreading.

PREFACE

THIS BOOK IS written in an urgent attempt to warn you about what I believe to be a global man-made health threat. When Thomas Edison began wiring New York City with a direct current electricity distribution system in the 1880s, he gave us the magic of electric light, heat, and power, but inadvertently opened a Pandora's Box of unimaginable illness and death.

There is a high likelihood that most of the twentieth century "diseases of civilization," including cardiovascular disease, malignant neoplasms (cancer), diabetes, and suicide, are not caused by lifestyle alone, but by certain physical aspects of electricity itself. The data to prove this has been available since 1930, but no one investigated it. Consequently, the "wars" on cancer and cardiovascular diseases are doomed to failure, because a critical etiologic factor has not been recognized. What's more, these very diseases are now increasing in the population in direct proportion to our increasing exposures to high-technology electrical devices.

The electrical part of this story begins with a childhood leukemia cluster centered in Rome, New York, that I studied in the 1960s. I didn't realize that the cluster was probably caused by radar exposure until many years later when Stanislaw Szmiegelski, a researcher in Poland, reported that radar and radio-exposed military personnel had high rates of leukemia and lymphoma (Szmiegelski 1996). In the United States, the emergence of childhood leukemia in the 1930s, and the spread of the age two-through-five-year peak for the major leukemia of childhood, common acute lymphoblastic leukemia, was strongly correlated with the gradual spread of electrification from urban into rural areas (Milham & Ossiander 2001). Even today, this childhood leukemia age peak does not appear in non-electrified areas like sub-Saharan Africa.

While conducting the childhood leukemia age peak study, a few adult cancers, including female breast cancer, also showed a strong correlation with residential electrification. At that time, I could not believe that 60-Hz magnetic fields could be responsible. A few years later, a 2004 newspaper article about a cancer cluster in teachers at the La Quinta Middle School in Southern California led me to conduct another study, which showed that high frequency voltage transients (called "dirty electricity" by the utility industry) was a potent universal carcinogen. Dirty electricity rides along on the sixty-cycle sine wave of alternating current (AC) power as high frequency voltage transients, between two and one hundred kilohertz. It also is increasingly found in ground currents returning to utility substations. They are caused by interruptions of current flow and by arcing and sparking. Dirty electricity can be present on electrified wires anywhere and probably has been on them since the beginning of electrification. Ambient dirty electricity couples capacitively to the human body and induces electrical currents in the body.

The La Quinta paper, published in 2008 (Milham & Morgan 2008), led to another study in 2009, "Historical evidence that electrification caused the twentieth century epidemic of disease of civilization" (Milham 2010), which motivated the writing of this book and my warning. This book will explain how a then seventy-two-year-old retired medical epidemiologist became involved with what turned out to be the most important, interesting, heart-breaking, and difficult series of studies of my long career.

The health and mortality effects of electrification happened so gradually, and on such a wide scale, that they went virtually un-noticed, and the major illnesses that can be attributed to them came to be considered "normal" diseases of modern civilization. Although major cities had electricity at the turn of the last century, it took until the mid-1950s for the last farms in the United States to be electrified. By 1940, more than 90 percent of all the residences in the northeastern United States and California were electrified. In 1940, almost all urban residents in the United States were, therefore, exposed to electromagnetic fields (EMFs) in their residences and at work, while rural residents were exposed to varying levels of EMFs, depending on the progress of rural electrification in their states. In 1940, only 28 percent of residences in Mississippi were electrified, while five other Southern states had less than 50 percent of residences electrified. Eleven states, mostly in the Northeast, had residential electrification rates above 90 percent. In the highly electrified Northeastern states and in California, urban and rural residents could have similar levels of EMF exposure, while in states with low levels of residential electrification, there were potentially great differences in EMF exposure between urban and rural residents. It wasn't until 1956 that these differences finally disappeared. What was already known by then, but not appreciated, was that urban death rates were much higher than rural rates for cardiovascular diseases, malignant neoplasms, diabetes and suicide in the 1930 and 1940 United States mortality data. In 1930, urban cancer death rates were 58.8 percent higher than rural cancer death rates. Rural death rates were significantly correlated with the level of residential electric service by state for most of the causes examined.

It is difficult to believe that mortality differences of this magnitude could go unexplained for more than seventy years after first being reported, and forty years after they had actually been noticed and commented upon. I suspect that in the early part of the twentieth century, nobody was looking for answers or knew how to properly frame the appropriate broad epidemiologic questions. By the time EMF epidemiology began in earnest in 1979, the entire population was exposed to EMFs. There was then simply no way to find an unexposed control group; therefore, all studies were potentially biased. Cohort studies, which follow groups of people forward in time, were by then using EMF-exposed population statistics to compute expected values, and case-control studies were comparing more exposed cases to less exposed controls.

By way of analogy, the mortality from lung cancer in two-pack-a-day smokers is more than twenty times that of non-smokers, but only three times that of one-pack-a-day smokers. Extending that analogy to EMFs, after 1956, the EMF equivalent of a non-smoker ceased to exist in the United States, with the exception of the small Amish population. The inescapable conclusion of these findings is that the twentieth century epidemic of the so-called diseases of civilization, including cardiovascular disease, cancer, diabetes and also suicide, was caused by electrification and the unique biological responses we have to it. A large proportion of these diseases may therefore be preventable.

We are an electrochemical soup at the cellular and organ level. Think of ECG (electrocardiogram), EEG (electroencephalogram), and EMG (electromyogram). We evolved in a complex EMF environment with an interplay of natural terrestrial and extra-terrestrial EMF sources from solar activity, cosmic rays, and geomagnetic activity. I believe that our evolutionary balance, developed over the millennia, has been severely disturbed and disrupted by man-made EMFs.

I believe that man-made EMFs, especially dirty electricity, are chronic stressors and are responsible for many of the disease patterns of electrified populations.

The very good news is that there are reasonable ways to eliminate or reduce this hazard if society chooses to do so, in ways that can make modern life far safer without requiring us to live in the dark. It took nearly fifty years of education and experience to place me in a position to really understand what the La Quinta school cancer data meant. Please join me in a trip back to Albany, New York, in 1932 as I explain how I got to here from there.

THE EARLY DAYS

I **WAS BORN IN** The Albany Hospital, in Albany, New York, on May 12, 1932. I would later attend medical school in the same old red brick building where I had been born, and my first son would also be born there.

My parents were Lebanese and Syrian, and both were Orthodox Christians. My dad, Sam, was actually born in Albany, but his parents had come to the United States in the early 1900s from Brummana, a small town near Beirut, Lebanon. My mother, Louise, was born in Tartous, Syria, and had immigrated as an infant with her parents. My mother graduated from high school, while my dad only finished the fourth grade, because he and his three siblings became trapped in Lebanon by World War I with their mother while visiting her parents. They were out of touch with my grandfather, Alex, for nearly four years, and were often hungry during the Turkish occupation of the Middle East. They all survived the war, but tragically, before I was born, my grandmother, Libby, died in a roll-over car accident while on a family trip. My dad was so traumatized by her death that he never drove a car.

I attended grades one through eight at Public School 26 in the west end of Albany, which was one hundred feet from our front door. The school was torn down a few years ago to make way for an office complex. This location was important, since I spent the seventh grade in bed with a respiratory infection but lost no school time, because the teachers were able to drop off my lessons after school on the way to their bus stop. After ninth grade at Philip Livingston Junior High School, I attended Albany High School, one public bus ride from home. Academically, the school skimmed off about twenty bright students out of the five hundred who attended and put them in special classes taught by system-wide department chairmen. The tangible benefit of the special high school education that a few of us received was that we could compete with the New York City kids, who had the benefit of attending special technical and science high schools for New York State scholarships. I won the German Prize and the Solid Geometry Prize at high school graduation. One day near the end of twelfth grade, I took an examination that won me a full tuition scholarship to any New York State college of my choice. I decided on Union College in Schenectady, New York, so I could live at home.

Union College was one of the oldest colleges in the country and was affiliated with a state university system that included the Albany Medical School, the Law School, and the Pharmacy School. With an early love of science, I enrolled in the pre-med program. The college had a lovely park-like campus, a fraternity system, and some exceptional teachers. With no family car, I hitchhiked the sixteen miles to and from college every weekday for four years. One of my pre-med classmates helped me find a weekend job at St. Clare's hospital in Schenectady working in the clinical laboratory on weekends for one dollar per hour plus room and board from Friday evening until Monday morning. Within a few months, I was single-handedly running the lab at night, drawing pre-op bloods, doing blood counts and chemistry profiles, taking blood samples for transfusions, typing, cross-matching, and doing bacteriology. The pathologist who oversaw the lab allowed me to assist in autopsies. He also loaned me his books, an old monocular Zeiss microscope, and a hematology slide collection that I could take home to study. Here, I got my first taste of the practice of medicine, and working in the lab gave me hands-on experience in clinical pathology.

In my senior year at Union, I took another test for a state scholarship and won full tuition to the Albany Medical College (AMC). At that time, I also won another small scholarship, the Fuller Chemistry Prize for Excellence in Chemistry, awarded to a Union graduate headed to AMC. I don't know if New York State still has full scholarship programs, but without them, my life would undoubtedly have taken a different course. Without such financial help, I would surely be doing something else today, and it's likely that none of the research described in this book would ever have taken place. This is something to keep in mind when government programs for education are cut.

ALBANY MEDICAL COLLEGE (AMC)

AFTER FOUR YEARS at a good liberal arts college, I thought that medical school was an intellectual step backward. It was trade school in the literal sense of the word. The scientific basis of medical education could be boiled down to the study of pathology, which is the physical, pathologic, and cellular basis of all disease. There was no formal course in medical history, but reading about those who went before us and how they solved the medical mysteries of their time inspired me then, and it still does. Most doctors arrive at their choice of practice by a process of elimination, balancing the positive and negative aspects of each specialty. Helping patients in their time of need and watching how the good doctors went about their craft was the best part of the process, and how I really learned. In terms of teaching medicine, I don't think that the apprentice system has ever been improved upon. I also think that it would be a very good experience for every med student to be an inpatient for a while.

When I attended the AMC medical school beginning in 1954, it was physically part of The Albany Hospital. The students had a locker room and had use of a lounge and the hospital library. There were fifty-four men but only one woman in my class. A Veteran's Hospital was across the street, and the New York State Laboratory and Health Department were within walking distance.

Although I'd already been acquainted with many aspects of medicine, including autopsies, while working at St. Clare's lab during college, the AMC years gave my fragmentary lab experience a medical perspective. The first two years of the curriculum were devoted to basic sciences like human anatomy, physiology, pharmacology, and pathology. Like all anatomy courses then, ours involved the yearlong process of dissecting a corpse. Our corpse was a male who looked like an unwrapped Egyptian mummy. It was hard to believe that this person was ever alive. The tissues were leathery hard and dark brown. The contrast between how the tissues looked in our corpse as compared to how they looked at autopsy or in a living person during surgery was striking.

I lived at home, worked nights at the New York State Laboratory as a biochemist, and in my four years, contrary to popular med-student stereotype, managed never to take a book home. I found plenty of time during the day to study and learned that medicine was in fact as much of an art as it was a science.

The third and fourth years were when we learned how to be doctors. We rotated through the various specialties, learning how to diagnose and treat actual patients. All the specialties were interesting, but in pondering what to do after medical school, I began a triage of what kind of medicine I *didn't* want to do.

Psychiatry was out. It had no pathologic foundation. Psychiatry also had treatments I found repugnant. I didn't like the use of electroconvulsive therapy back then and am sickened to see it making a comeback now. I also thought that psychoanalysis had no scientific basis.

My obstetrical rotation, on the other hand, was a happy service, because it involved young women doing what comes naturally. Every delivery of a baby was exciting to me, and I enjoyed helping the mothers get through their labor. There was a lot of induced labor back then, mostly for the convenience of the obstetricians. I also thought that the cesarean section rates were way too high. Though rewarding, I crossed obstetrics off my list because it was too much of the same thing, with very little diagnostic challenge.

I did find pediatrics very interesting, but heartbreaking. I couldn't handle watching an innocent child die. Pediatric oncology was the worst, because in those years, all childhood leukemia was fatal. At the time, we used Nelson's *Textbook of Pediatrics*. In each section was a brief discussion of the epidemiology of a given condition. I quickly discovered that the epidemiology of a disease, and how it was understood, was a lot more interesting to me than how to actually treat a disease. After more independent reading, especially about how the acute communicable diseases of childhood were understood, I suspected I had found my medical niche.

We also had a surgical rotation. I liked the diagnostic part of surgery, because you could use your clinical, detective, and laboratory skills to reach a diagnosis and could find out directly if you were accurate.

However, I saw some things at the operating table and on the wards that made me wonder how anyone could survive hospitalization. Eventually, I crossed surgery off my list, because only the diagnostic part was interesting to me.

I did like internal medicine. The teacher I admired most and tried to emulate was a tall, soft-spoken gentleman named Gilbert Beebe. He said many times that listening to the patient was critical and that the patient, if properly questioned, would always tell you the diagnosis. That wisdom would later come in handy with my work on dirty electricity and with people who had become hypersensitive to electromagnetic fields. Their doctors typically don't believe them, but they are being given the diagnosis.

Throughout the various rotations, it had become increasingly obvious to me that I was a lot more interested in understanding what brought people into the hospital than in how to treat them. My path toward epidemiology became stronger by the day.

My personal life had also taken some interesting turns during medical school. Between my sophomore and junior years at AMC, I taught archery at a summer camp in Vermont. There, I met an attractive school teacher named Lorna Galbraith, who taught tennis and had a car. We married a year later in 1956 and had our first of three children, a son named Richard, in 1957. By the time I graduated, I was a husband and a father. A daughter, Suzanne, was born in 1959, during my internship in Boston; and a son, Sam, was born in 1961, while I was at Johns Hopkins in Baltimore getting a masters degree in Public Health.

One recollection I have about medical school graduation was that as each of us walked across the stage to get the piece of paper we had worked so hard for, I asked myself which of my fifty-four classmates could I trust with the life of my kids. Sadly, I came up with only six names. It's probably a good thing that I have absolutely no say in who gets to practice medicine.

INTERNSHIP

INTERNED AT A U.S. Public Health Service (PHS) Hospital in Boston called Brighton Marine, which is no longer there. The PHS looked after fishermen, U.S. Coast Guard personnel, and other military dependents. With no money and a growing family, I needed a paycheck to stay afloat during internship and residency, so my internship options were few. PHS offered a small but livable stipend for interns and residents, unlike most private hospital training programs.

We packed up our Studebaker Commander V8 and headed east to Boston. By then, Lorna was pregnant with our daughter Suzanne. I was the last intern to arrive, and learned that I would immediately begin a two-month obstetric rotation at St. Margaret's Hospital in Dorchester. After that, I would have two months at Boston City Hospital learning pediatrics, before coming back to the PHS hospital.

It was a difficult year. Lorna had a first trimester bleed and was put to bed, and Richie, our toddler, had recurring streptococcal throat infections.

St. Margaret's was a very busy place. Boston was a Catholic city.

With only the rhythm method for birth control, Catholics had lots of babies. The work was forty-eight hours on call, followed by twenty-four hours off. In busy periods, it was hard to get any sleep.

We had to do circumcisions every day, using a crushing clamp. I still can hear the cries of those little guys. While doing the circumcisions, I noticed that a few of the babies were mildly jaundiced. After a little record checking, I found that they were all delivered by one obstetrical group that used a lot of heavy-duty narcotics to knock the mothers out to the point that they didn't remember anything about their labor or delivery. I think the technique was called "twilight sleep." It was bad for the mothers and especially bad for the babies. The mothers would thrash about on their cots, sometimes soiling themselves, and pulling out IVs. The babies were born depressed and difficult to resuscitate. They also were jaundiced. When I reported my findings to the hospital director, he told me to keep my mouth shut or they would terminate my training. What an auspicious start to my research career. It was a hectic, exhausting, exciting, and rewarding time, but two months of it was enough.

I transitioned right from obstetrics to pediatrics at Boston City Hospital. Dr. Sidney Gellis was chief of service then, and I think I saw him once. I had no ward responsibilities, but worked the evening shift in the pediatric emergency room, as well as the following afternoon in specialty clinics. When I saw interesting cases at night, I could refer them to the clinics the next day, when there were senior pediatricians around for consultation. The emergency room (ER) was a crowded, busy, chaotic place. Assuming all that responsibility right out of medical school was something I never anticipated. There were four experienced nurses who showed me the ropes, and because of them, I survived. I learned in all my new clinical assignments that it was critical to get the head nurse, oftentimes someone who had been there for twenty years, on my side. To my misfortune, I started my pediatric rotation in the middle of an epidemic of aseptic meningitis. The kids would come in with fever, stiff neck, and vomiting. There was no effective treatment, since the disease was caused by a number of different viruses. Every meningitis patient needed a spinal tap, and I had to do the lab work myself. During the epidemic, this meant one or two cases per evening, and the spinal fluid always ran clear, indicating that it was a viral meningitis. One night before the night shift took over, however, another meningitis case showed up. I was tempted to bypass the spinal tap, but a little inner voice said, "What if …?" That one turned out to be bacterial meningitis, the only such case I saw in my two months. Instead of a clear spinal tap, pus ran out of the needle. Any delay in treatment could have been fatal. The children in the ER were wonderful, and I loved treating them.

After pediatrics, I returned to PHS to begin rotations in medicine, surgery, orthopedics, and the emergency room. I had been away from PHS for four months at St. Margaret's and Boston City Hospital, and was disappointed to learn that all the specialized clinical residencies for the next year had been assigned. That left me the choice of working in the hospital ER for another year and then getting a clinical residency, or taking a public health residency, which was still open. I decided on the latter.

Working on the wards was very interesting, because this was real "doctoring." I got to admit the patients, decide what was wrong with them, work out a treatment program, check on their progress, and finally send them home if everything worked out well. There were some memorable cases that taught valuable lessons.

A young fisherman was admitted with swollen lymph nodes, fever, tiredness, and a general malaise (feeling lousy). His working diagnosis was some sort of lymphoma or Hodgkin's disease, and I remember the tears and the anguish when his wife and family got the news. But something about the diagnosis just didn't ring true. There was no weight loss, and the onset of his illness was sudden. Usually, cancers develop slowly, with symptoms evolving over time. The first indication of a lymphoma is often a painless swollen lymph node in the neck.

One condition that causes a similar presentation is a bacterial infection called cat scratch disease or cat scratch fever. The intern and resident on his floor had asked the patient on a number of occasions about being "scratched" by a cat. He said he had not. One of the residents also agreed that something about the case didn't seem right, so we decided to reread the literature on cat scratch disease, which mentioned that it can also be caused by a cat bite. When asked the right question (just like my medical school professor, Dr. Beebe, correctly taught us), the patient remembered having been bitten by a kitten. His skin test was positive for cat scratch disease, and he was discharged to recover at home.

The case that taught me a most important lesson about listening to patients was a seventy-year-old woman who had been plaguing the hospital staff for years with an assortment of vague complaints. Patients who had been seen at the hospital before arrived with their medical records. This woman arrived with a file that weighed almost as much as she did. I waded through her records, and it became clear that in recent years, no one had taken her seriously due of her past history of frequent complaints. I decided that she was going to get the same exam as any other admission. It turned out that she had a fatal, inoperative colon cancer, and she died before I finished my internship. The moral is that even seeming malingerers get sick and die, and they shouldn't be dismissed out-of-hand.

One very difficult two-week period involved a tanker collision in Narragansett Bay off Providence, Rhode Island. A number of merchant seamen were severely burned when they were forced to jump into water aflame with burning oil. I learned how difficult burns are to manage, for both patients and the people treating them. I can see why special burn treatment centers have evolved over time. The care of burn patients is a science unto itself.

I was sent in a small ambulance to transport about a dozen burn victims, two at a time, from Providence to the PHS Hospital in Boston. I remember the smell of burnt flesh and the cries of pain every time the ambulance hit a bump. From the way that "just" a dozen burn cases tied up practically our entire hospital, it was obvious that there would never be a way to handle large numbers of people injured in a major natural or man-made disaster.

Another interesting case involved me, and illustrated the power of therapeutic electromagnetic fields on the body. During medical school I had been plagued with a nagging, painful case of plantar warts on the soles of both feet near the toes, and on the ball of my right big toe. I also had a small ordinary wart near my knee. The plantar warts had been whittled, frozen, and burned, but they kept coming back. Then one slow night in the ER, I used an electrocautery needle to burn off the small wart on my leg above the knee. It bubbled when the current was delivered, after which I just brushed it off with the side of a scalpel. I applied the current to the empty crater once more with the needle, put a band-aid over it, and went back to work. Amazingly, in less than a week, all my plantar warts were shrinking, even though they had not been treated. In two weeks they were completely gone, never to return. My guess is that somehow I'd immunized myself to the wart protein when I burned off the little wart, or that some aspect of the electric current had affected my immune system. This would be a prescient guess, years before the biological effects of low-level EMFs had been studied.

My pull toward epidemiology had only deepened throughout my internship. We saw so many of the same diseases that I had become profoundly interested in backing the process up to the prevention stage, not just routine treatment after-the-fact. This seemed like only common sense. The great public health triumphs of the last two centuries were achieved by understanding what was causing disease and then intervening. Prevention seemed a lot more important to me than treatment.

The epidemic intestinal diseases, including cholera and typhoid, were prevented by confining human waste inside sewage pipes and by providing clean drinking water. Other nasty diseases like small pox, polio, diphtheria, and the acute communicable diseases of childhood had largely been tamed through understanding immunity and using it to prevent them. I read with great interest the story of John Snow, who unraveled how cholera spread through contaminated water supplies in London. This was long before he or anyone knew about the cholera bacillus. I came to think that these same techniques might work for understanding and preventing our modern plagues too. I really started to believe that good epidemiology as exemplified by Peter Panum's work on measles and Joseph Goldberger's work on pellagra (a serious, multi-system disease caused by a dietary niacin and protein deficiency), could one day prevent cancer, diabetes, and other contemporary diseases of civilization. I still feel that way fifty years later. In fact, I believe it more than ever after our work linked cancer to dirty electricity.

RESIDENCY

WHEN IT CAME time to decide what further medical training to undertake, I chose a public health residency. The other clinical residency slots were already filled, and my only option was to spend another year in the ER while waiting for a clinical residency. I felt it was time to move on.

Lorna and I now had two children. My daughter, Suzanne Elaine Milham, had been born in The Boston Lying-In Hospital in 1959. She came home with streptoccal disease and had to have a couple of her tiny little fingers cut open to drain the pus.

I was assigned to the Monroe County Health Department in Rochester, New York, for my first year of public health residency. We rented a house in Honeyoye Falls, New York, and I drove sixteen miles to work each day.

The first week on the new job, I came close to quitting. At the health department, I was assigned an antique desk in a cavernous loft in an old school building with no one in sight and nothing to do. This was in stark contrast to the busy, exciting internship in Boston that was still fresh in my mind. Like all medical school graduates, I had the option of going into private practice. It was a tempting idea, but I didn't do it.

I got busy running a well-baby clinic and a venereal disease clinic at the health department, but my first real practical epidemiology experience involved the Sabin oral polio vaccine trials in Rochester that the health department participated in. We would visit schools and drop the vaccine on small, extended tongues. The kids looked like baby birds with their mouths open and their heads tilted back to receive the vaccine from an eyedropper. For me, the most interesting part of the trial was gathering and evaluating complaints about reactions to the vaccine. I traveled all over Rochester and saw some very sick children, but none of them had a vaccine-related problem. I saw my first and only case of acute glomerulonephritis, a life-threatening kidney disease with characteristic smoky urine as one sign. I had to convince the parents that this disease was not caused by the vaccine.

Now I was conducting hands-on epidemiology, working with whole segments of the population. Simply put, epidemiology is the study of the distribution of diseases in populations. Epidemiologists attempt to determine disease causation by studying attributes associated with disease. They use some intuitively simple study designs, which can be very difficult to execute at times. Epidemiology, for instance, proved that cigarette smoking caused lung cancer. In a case-control study, people with lung cancer (cases) are compared with people without lung cancer (controls) for a history of cigarette smoking (exposure). In a cohort study, populations are followed in time. In a large cohort study involving British physicians, a smoking history was taken, and the cohort was followed. It was found that two-pack-a-day smokers had a twenty-fold greater lung cancer mortality rate than physicians who never smoked; and three times the lung cancer mortality of one-pack-a-day smokers.

While at Monroe County, I designed and conducted my first small study in a high school population to determine diphtheria immunity levels, and how they related to previous immunization. It's remarkable that most of the physicians of my generation have never seen a case of diphtheria. This was testimony to the success of our early public health efforts and the efficacy of using the DPT vaccine. Diphtheria became rare in my lifetime.

Near the end of my time at Monroe County, some family problems arose unexpectedly. We were having dinner in the basement of our rented house, since the upstairs was too hot during the summer months. My mother was visiting and somehow forgot that a gas burner was left turned on under an oil fryer in the kitchen upstairs. I heard a crackling sound as the wood cabinets above the stove caught fire. I dashed upstairs, shut the gas off, grabbed the pot with the burning oil, and headed toward the outside door. But when I opened the door, the wind blew the flames back into my face and scorched the part of my hand not covered by the potholder. Now I really understood what those merchant mariners felt back in Rhode Island when they were so badly burned in that tanker fire. In the end, I wasn't burned too badly, but I still have a small burn scar on my right hand. The most troubling aspect was that the landlady asked us to leave.

We found a small summer cabin on Canandaigua Lake for the six weeks I had left at the health department. A few days after we moved in, my infant daughter, Suzanne, became quite ill with a high fever and a bright red throat. I took a throat culture and drove to town to get some penicillin. The throat culture grew out beta hemolytic streptococcus. Further investigation revealed that the same bacteria grew out of the milk we were drinking. Additional checking with neighbors proved that our little lake community was in the middle of a milk-borne strep epidemic. Since the store that sold the milk was in a different county from the laboratory that would routinely handle cultures from our community, no one in a position of authority ever tracked down the source of the contamination. This was a clear example of a public health failure, and I've been witness to many others.

SCHOOL OF PUBLIC HEALTH

At the end of my year-long residency in Rochester, we packed up again and headed to Maryland, where I started my Master of Public Health year at Johns Hopkins University School of Public Health. We rented a house in Glen Burnie.

At Hopkins, I had a new director from the Public Health Service named Leslie Knott, but I only saw him at the beginning of the year and at the end.

The class was made up of young-to-middle-aged physicians and nurses from all over the world. Interacting with my peers from different cultures was among the best parts of the Hopkins experience. One of my classmates, a burly Yugoslav named Fred Zerzavy, would eventually deliver our third and last child, Samuel Glen Milham, while we were in Baltimore.

There were some outstanding faculty members, such as Abraham Lilienfeld, who taught some of the epidemiology course. He would later be in charge of studying the effects of microwave radiation on U.S. State Department personnel at our embassy in Moscow, Russia. Dr. Lilienfeld began his first lecture with a simple statement: "The problem with epidemiology is the controls." No truer words about the field were ever spoken. Identifying and characterizing cases is easy, but getting proper controls is often difficult, even impossible. By way of example, recent cell phone/brain cancer case-control studies had control participation rates of only about 50 percent. Now, there are many reasons why people who agree to participate might be different from people who refuse, including concerns about privacy, time constraints, and individual temperaments. Unless a researcher can somehow prove that the participating controls are representative of a broad, accepted, and recognized control group, all such work is ripe for dismissal. This is one of the inherent, vexing aspects of epidemiology. But despite this, epidemiology is the only way to capture data for human populations.

When I first started publishing papers in the 1960s, we were held to a much higher standard than is the case today. If a researcher didn't have 90 percent control participation, he couldn't get a study published. Philip Sartwell, then editor of the *American Journal of Epidemiology*, presided over the Hopkins Epidemiology Lab with assorted assistants who were also well known in the field. Over the years that he was editor, Dr. Sartwell sent me some important papers to review, including one authored by Nancy Wertheimer, who, with Ed Leeper, started modern EMF epidemiology with a 1979 paper on residential EMF and childhood cancer in Denver, Colorado.

Shortly before I was to finish the MPH degree and move on in the residency, I was visited by Dr. Leslie Knott. The exchange did not go well. He offered me a couple of residency positions, one in California and another somewhere I can no longer recall. I told him that I'd be happy in any position if he could guarantee me at least two years in one place. This annual moving was taking its toll on me and the family. He blew up and said that being a member of the PHS was like being in the Army, and that he could make no such guarantee. He then asked me to evaluate my year at the school of public health for him. I assumed that he wanted the truth, so I detailed the strong and the weak points of the year. He erupted again and accused me of being a malcontent. "Hopkins is a great school. You don't deserve to go to a school like Hopkins!"

I thought his tirade was the end of it, but a week later I got a late night phone call from Dr. Knott's secretary. "Dr. Milham, he is going to have you drafted." she said. "You need to spend two more weeks at Hopkins to fulfill your draft requirement." The school allowed me put in two more weeks (at that time, public health service counted toward military service), after which I signed up for a third residency year at the New York State Department of Health in Albany.

It was good to be back in my hometown with family support and lots of friends both in and out of the field. I was close to my old school, Albany Medical, and its library, plus I acquired a faculty position in pediatrics there.

My residency was directed by William Haddon, who later went on to head the Federal Traffic Safety Commission. By mutual agreement, Haddon and I decided that my residency would be better served working with the head of vital statistics, a tall, bright, slightly rumpled Ph.D. statistician named Alan Gittlesohn, who was glad to have me. Alan had room for my desk and even helped me push it down the hall from Dr. Haddon's office to his. We began a four-year friendship and research relationship from which we would both learn a lot.

After this third year of public health residency, I signed on with the New York State Department of Health. The move was just down the hall to a new office, working for Robert Korns. Alan Gittlesohn and I continued our work together on special projects. I finally felt somewhat settled.

NEW YORK STATE DEPARTMENT HEALTH

MY FIRST JOINT epidemiology project with Alan Gittlesohn was to study congenital defects reported in New York State birth and stillbirth records. The health department covered the upstate counties and included the Long Island counties of Nassau and Suffolk. New York City had its own government and vital records system.

Before jumping into the study, I tried to assess how well the congenital defects were reported in the state birth and stillbirth certificates. The state had a system that paid for the surgical repair of certain birth defects, so I obtained case rosters from them. I also visited the three upstate teaching hospitals at Albany, Rochester, and Buffalo and had them pull all their charts for certain diagnostic categories of defects. When I compared the records, it was apparent the central nervous system (CNS) defects called anencephalus (absence of a major part of the brain) and spina bifida (spinal column defect) were well reported in the birth and stillbirth records (more than 80 percent). Cleft lip and palate was also well recorded (75 percent). So we had some good data to work with.

We wanted to study the incidence of these devastating birth defects over time to search for etiologic clues. In epidemiology, incidence is the number of new cases per year per unit of population at risk. In mathematical parlance, incidence is just a fraction with the case count in the numerator and the population at risk in the denominator. In this case, the population at risk was the population of births. Fortunately, New York State has a large population. We were able to study more than 8,000 births with central nervous system defects in a population of more than 2.5 million births. The major finding of the study confirmed that more female than male newborns suffer from these defects, and we also observed a 50 percent decline in incidence over the years 1945–1959. Since we did this study (Gittlesohn & Milham 1962), it has been learned that dietary supplementation with folic acid before and during pregnancy can prevent 70 percent of these defects. I think that in the United States, folic acid was being added to certain foods after World War II. This would account for the declining incidence of central nervous system defects that we found.

Although nearly all congenital defects have a familial component, very few studies had determined the repeat frequency of CNS defects in families. I collected data on 139 families with a child born with either anencephalus or spina bifida, and found that ten of their 308 brothers and sisters had a similar defect (Milham 1962). This is about sixteen times higher than what would be expected.

CHILDHOOD LEUKEMIA

My first foray into investigating leukemia incidence in populations began in the early 1960s when the British journal, *The Lancet,* published a query asking for information about leukemia clusters. It would turn out to be an area that would hook my interest for decades to come.

This was the pre-computer era. To do the study, I used a card-counting sorter to sort death record punch cards for all the children dying under age ten of leukemia in the years 1948–1960 by county. Using 1950 and 1960 census population for denominators, I found that in 1950, New York state had a childhood leukemia death rate of 4.3 per 100,000 people, but Oneida County alone had an elevated rate of 19.9 per 100,000 people. Between 1950 and 1951, Rome, Oneida County, New York, had a greatly elevated childhood leukemia rate of 36.1 per 100,000 people. I verified the Rome, New York, diagnoses through both pathological information and interviews with the next of kin. One father told me that his wife had also died of acute leukemia at age twenty-eight, one year after their son had died. Unfortunately, at the time, I could find no common denominator to explain the cases (Milham 1963). In 1996, when researcher Stanislaw Szmiegelski showed that Polish military personnel working with radar had an increased incidence of leukemia, I remembered that many of the homes I visited in Rome, New York, were in line-of-sight with the large rotating radar arrays of Griffith Air Force Base. This offered further support of my suspicion that leukemia might be caused by environmental electrical exposure.

In reviewing the epidemiology literature about childhood leukemia, I came across a remarkable 1961 paper by Michael Court-Brown and Sir Richard Doll showing that a peak in childhood leukemia at ages two through four had emerged anew in the United Kingdom in the 1920s; and in the United States white population in the 1930s (Court-Brown & Doll 1961). Childhood leukemia was then a rapidly and completely fatal disease, which made death records an efficient way to count cases, since records would be recent and would count all cases. I reasoned that if we could find something that had changed at about the time the peaks appeared, we could solve the childhood leukemia mystery. It took about forty years, but I finally was able to show that the childhood leukemia peak was caused by some facet of residential electrification (Milham & Ossiander 2001). More about that later.

CONGENITAL DEFECT SURVEILLANCE

Shortly after starting work at the New York State Department of Health, I thought it would be interesting to do a real-time analysis of congenital defects. At the time, all of the vital records from the upstate population passed over just a few desks in Albany each month to be manually coded before being keypunched and stored. Very detailed annual reports were generated from this information, with a two or three year delay. I wanted real-time analysis so I set up a simple birth record review system. The problem was that rare events like congenital defects and childhood leukemia are so dispersed in the population, that no single doctor would be likely to notice an excess incidence in his practice. For birth defects surveillance, I asked the coders to make me a copy of all the birth and stillbirth records that listed a congenital defect, and did a monthly analysis of malformation incidence. I used a previous year of malformation experience for a baseline.

A few months after starting the system, there were a higher number of infants born with missing arms or legs than expected. I called the mothers, and found that almost all of them had been out of the country and had taken thalidomide. Before I had a chance to complete the investigation, the phocomelia-thalidomide connection appeared in the medical literature. It was thrilling to know that my simple system picked up the association in its early stages.

My system had another rigorous test when a pandemic of German measles swept the state. The virus can cause eye defects in fetuses if the mother is infected. The system picked up the eye defects with no problem, and the stillbirth rate increased and the live birth rate decreased, showing that a lot of the impact of the pandemic on births happened early in pregnancy before vital record ascertainment was possible.

When I left the health department four years later, the surveillance system left with me, since no one at the department was interested in maintaining it. When I joined the Washington State Department of

Health in 1968, after a short tour in Hawaii, one of my first projects was to set up a similar birth record surveillance system.

In 1970–71, eleven infants were born in Washington State with scalp defects reported on their birth records (Milham & Elledge 1972). The lesions were single, circular, punched-out ulcer-like midline defects at the top or over the back of the head. A query of the mothers revealed that two of them had taken methimazole (Tapazole, manufactured by Eli Lilly) during pregnancy for hyperthyroidism. One of the mothers delivered fraternal twins, both of whom had the defect.

I published a note in the journal *Teratology* asking if others had noticed this association. A number of physicians around the world had seen the same scalp defects in the infants of women taking methimazole, and one physician reported that a mother taking carbimazole had had an infant with a scalp defect. Carbimazole is metabolized by the body into methimazole, strengthening the association. Additionally, some of these infants were reported to have an umbilical cord defect called patent urachus.

I reported my findings to the drug company and to the Food and Drug Administration. A warning was added to the drug insert, and gradually other overactive-thyroid drugs replaced methimazole. I think this was the first, and possibly only, time that routine birth record malformation surveillance identified a human teratogen. This was probably the only way the association could have been made.

TWIN STUDIES

Gittlesohn and I also conducted a large study of twins (Gittlesohn & Milham 1965) in New York State, linking the birth records of single births into a twins pair file. We wound up with 21,128 twin pairs among all births between 1950 and 1960. There are two types of twins: identical or monozygotic (one egg), and fraternal or dizygotic (two egg). In our study, we examined fetal survivorship, sex, birth-weight, zygocity (one egg or two eggs), maternal age, and birth rank. To me, our most interesting finding was that for any maternal age, the fraternal twinning rate increased systematically with birth rank. This held true for any maternal age with the number of previous pregnancies determining the probability that a woman would have fraternal twins. In other words, the more previous pregnancies a woman had, the higher her fraternal twinning rate became.

I knew that two-egg twinning was the result of polyovulation (more than one egg). A drug called clomiphene, given to infertile mothers, caused polyovulation and consequently many multiple births. Ovulation is under the control of the pituitary gonadotrophic hormones, follicle stimulating hormone (FSH), and leutinizing hormone (LH). I suspected that the pituitary gland had to be somehow responsible for the birth rank/fraternal twinning relationship.

My aha! moment came when I found a reference in my medical school pathology textbook showing that the weight of the pituitary gland of women increases with successive pregnancies. In 1964, I published a medical hypothesis in *The Lancet*, stating that fraternal twinning was caused by multiple ovulation, which in turn was caused by excessive production of the pituitary gonadotrophic hormones FSH and LH (Milham 1964). Years later, when FSH and LH levels could be accurately determined, my hypothesis was confirmed. This turns out to be true for others species, too. The litter size in dogs and cats increases with the number of litters.

LEUKEMIA IN HUSBANDS AND WIVES

In the early 1960s, while visiting at the Albany Medical College, a hematologist complained he was having trouble treating his leukemia patients. There was some scientific speculation at the time that human leukemia might be caused by a virus and could be spread by personal contact. This speculation created problems among families and clinicians alike. He asked me if there was any way to examine the question. Three years later, I had his answer.

I studied adult leukemia in husbands and wives, selecting all leukemia deaths in the years 1951–1961 where the marital status of the deceased was stated as "widow" or "widower." Of the 1,241 cases, a spouse's death certificate was located for 876 of them. Out-of-state death was the major reason for failure to find a spouse's death record. A matched death record control for each spouse was selected. There were seven leukemia cases in the spouses of the cases and five in the controls. The cause of death distribution of the spouses and their controls were similar. I was happy to conclude that adult leukemia is not contagious in the usual sense (Milham 1965).

HODGKIN'S DISEASE IN WOODWORKERS

My last study in New York before I left for Hawaii set the course for the rest of my career, by showing me the power of occupational studies. One of my medical school professors was dying of stomach cancer, and his college-age daughter was back in Albany to be closer to him. She volunteered to help in a study, which she selected from my "to do" list. We reviewed the death records of 1,549 white males who had died of Hodgkin's disease and their matched controls, and found a two-to-one excess of woodworking or wood-exposure occupations in the Hodgkin's cases (Milham & Hesser 1967). In Hodgkin's disease, the exposure was to wood. How many other cancers might actually be attributable to other environmental or occupational factors?

UNIVERSITY OF HAWAII AND THE WASHINGTON STATE DEPARTMENT OF HEALTH

SHORTLY AFTER COMPLETING the Hodgkin's disease study, I headed to Honolulu to take a job teaching epidemiology in the school of public health and the new medical school at the University of Hawaii.

The Hawaii job seemed ideal, but my oldest son, Richie, became chronically ill there with a low-grade fever and allergies. On a trip back to the West Coast during our first summer, he was well, only to become ill again on our return to Hawaii. His health problems would be a major factor in our relatively short stay.

I received a small grant from the Hawaii Cancer Society to replicate the Hodgkin's disease study in the states of Oregon and Washington, where they have lots of woodworkers. West Coast woodworkers also had a Hodgkin's disease excess, so the New York Hodgkin's woodworker findings were published in *The Lancet* (Milham & Hesser 1967). While visiting the Washington State Department of Health in Olympia, Washington, during the course of the study, I found an advertisement for the job that would take us back to the West Coast in May 1968. It would finally be our last job move.

My brief time in Hawaii was an eye-opener for me. The people I worked with were smart, and there always was something to be learned in other disciplines in the university environment. I loved teaching, and had a wonderful friend and boss in Bob Worth, head of the epidemiology department. We had moved from upstate New York in the dead of winter to a sunny and warm tropical paradise. The week before we left, I bundled the kids up on a sparkling cold winter's day to go sledding in the Helderberg Mountains near Albany, with a thermos of hot chocolate and cookies. Soon after that, we were in the tropics.

We spent a couple of nights at the Kaimana Beach Hotel while buying two old cars and renting a nice house in Hawaii Kai near Haunauma Bay. My children were the only Caucasians in the Koko Head elementary school and have been racially colorblind ever since. Lorna took a job teaching at a local college. This was during the Vietnam War, and my time at Tripler Army Hospital still haunts me. I saw many eighteen-year-old boys returning from that wretched war with permanent injuries that included paraplegia, double amputation, blindness, and brain damage.

WASHINGTON STATE HEALTH DEPARTMENT

We arrived in Olympia, Washington, in May 1968 to begin my job at the State Department of Health. I would eventually retire from there in 1992.

All the moving around had taken its toll. In 1971, Lorna and I would divorce. The children, by then ages eight, ten, and twelve, stayed with me in Washington State. Lorna went to California. In time, after the kids were off to college, I had other close relationships with smart, good women, one of whom died of a malignant melanoma at age forty-eight in 1990 after we had been together for ten years. She had unfortunately used tanning beds. Tanning bed use has subsequently been linked to malignant melanoma. Some jurisdictions are attempting to restrict or outlaw tanning bed use.

OCCUPATIONAL MORTALITY STUDIES

The New York Hodgkin's study and its association with the woodworking occupations lead me to think that a lot of unknown occupational disease associations might turn up through a systematic analysis of the occupational and cause-of-death information contained in the Washington State death records. The intuitive appeal of occupational diseases is that they are, by definition, preventable. Disease prevention is every bit as important as treatment after disease appears. However, prevention is rarely noticed, funded, or rewarded.

I did some small case-control studies of single cancers and single occupations. In the course of these studies, I developed some computer software to automate the studies. One program is still in use at the Hutchinson Cancer Center in Seattle, some thirty-five years later. Soon I concluded that the piecemeal approach made no sense, and decided to examine all occupations and all causes of death. This approach was used in the United Kingdom by the Registrar General and in the United States by the National Center for Health Statistics. Both studies examined many causes of death in many occupations. The Registrar General's study is done every ten years, while the U.S. study was done only once. The United States study used a proportionate mortality analysis, but used an age sixty-four cut off, so it missed most of the deaths. Before jumping into a massive, long-term commitment to occupational coding of deaths, I thought it was smart to see just how comprehensive and accurate the occupational and industry information was on the death record.

I was the thesis advisor for a doctoral student named Gerald Petersen at the University of Washington School of Public Health.

Petersen interviewed the next-of-kin of men who died from Hodgkin's disease and found that the death certificate statement for occupation was identical to that obtained in the interview in 75 percent of cases, and gave a related occupation in 10 percent of cases. We found the same results for men who died from bladder cancer.

I had also conducted a large population-based mortality study of members of the AFL-CIO United Brotherhood of Carpenters and Joiners of America (Milham 1974) and found that their death certificate occupational statements nearly always listed a trade covered by the union. With the assurance that the occupational information was quite good, my secretary and I spent a year abstracting and coding adult male Washington State deaths. We used a modification of the United States Census Bureau's Occupational code, adding codes for occupations specific to Washington State.

The National Institute for Occupational Safety and Health (NIOSH) published our first report in 1976, covering male deaths from 1950 to 1971 (Milham 1976). A first update was published in 1983 covering the years from 1950 to 1979; and before I retired, we reported an online analysis of 588,090 white male deaths from 1950 to 1989, as well as 88,071 white female deaths from 1974 to 1989.

With a study spanning forty years of deaths, three different International Classifications of Disease (ICD) codes for cause of death were in use. I remember living with the three different colored ICD books for about three months to make sure that we had an accurate correspondence between the ninth, eighth, and seventh ICD codes. (You can go to https://fortress.wa.gov/doh/occmort/ to look up your occupation and see its occupational mortality pattern). We used an age, sex, and year-of-death proportionate mortality analysis program, which calculates expected deaths based on the percent distribution of a given cause of death of all deaths in all occupations. A detailed cause of death analysis is available for 161 causes of death in 219 occupational categories for men, and sixty-eight occupational categories for women.

We knew we were on the right track when all the expected

occupational mortality associations were present in our data analysis: pilots die in plane crashes, loggers are struck by falling objects (trees), roofers fall, and power linemen are electrocuted. Most of the occupational mortality associations in the scientific literature were also present. Hard rock miners die of silicosis, asbestos and insulation workers die of lung cancer and pleural mesothelioma, and funeral directors and embalmers have increased leukemia mortality. The point of the study was to look for new occupational mortality associations, and a number of them emerged.

Below are some of the new occupational mortality associations we discovered.

MULTIPLE MYELOMA IN HANFORD PROJECT WORKERS

Workers at the Hanford Project, where weapons-grade plutonium for our nuclear weapons was made, had a small increased number of multiple myeloma deaths. I contacted Dr. Thomas Mancuso of the University of Pittsburg, who headed a government-supported study of the health of these workers, and gave him my Hanford workers file. I later sat on an advisory group that oversaw the health projects at Hanford. The myeloma excess held up, and it still shows up in multi-site analyses and is related to ionizing radiation exposure in workers measured with film badges.

To make sure that the populations living near the Hanford facility were not being adversely impacted by ionizing radiation from the site, I studied the local populations through their vital records, school records, and parental employment records. When the government admitted intentionally releasing radio iodine (RI) from the site, I spent a lot of time visiting the "downwinder" families, and searching for any evidence of illness or death. Happily, I found nothing, but that wasn't what some people wanted to hear. Since radio iodine is a thyroid seeker, the downwinders expected that they would have increased thyroid cancer and thyroid disease. Due to public pressure, the government then funded a large project to search for thyroid morbidity. It also found nothing. Large populations have been given RI therapeutically and diagnostically over many years, with no evidence that it causes thyroid disease or cancer.

LEUKEMIA AND LYMPHOMA IN ALUMINUM REDUCTION PLANT WORKERS

There were six aluminum reduction plants in Washington State where aluminum is made from aluminum ore (alumina or bauxite). The plants were located in the state during World War II because of its cheap hydroelectric power at that time.

The process involves the electrolytic reduction of alumina in an anode, cathode process and uses enormous amounts of electrical power. These workers had increased mortality rates from leukemia and non-Hodgkin's lymphoma in the Washington State occupational mortality study. I telephoned the medical directors of the companies, and was dismissed by all of them except for the medical director of Kaiser Aluminum, Dr. James Hughes. Kaiser had a reduction plant in Tacoma, and another in Mead near Spokane. They also had an aluminum rolling mill in Spokane. I explained my findings, and when I suggested a cohort study in the Mead worker population, Dr. Hughes agreed (Milham 1979).

The plant was built in 1946, and 25 percent of the workers were hired that year. There was a rumor that some of the worker records had been purged, but I found a proverbial "little old lady" in the personnel department who had saved a three by five card on every worker who had ever worked at the plant. These cards had all the information needed to do a good study. Using Richard Monson's modified life table computer program for analysis, I found increased mortality due to lymphatic and hematopoietic (blood) cancers, including leukemia, non-Hodgkin's lymphoma, benign brain tumors, pancreatic cancer, and pulmonary emphysema.

The Mead plant was a pre-bake-type plant, in that the anodes used in the electrolytic reduction process were baked before being deployed. This process minimizes air pollution in the potrooms, the place where the electrolytic reduction takes place. The Tacoma plant was a horizontal stud Soderberg-type plant, where the anodes were baked in place in the potrooms. The anodes and cathodes were made of coke with a coal tar pitch binder. Coal tar pitch gives off volatile substances when burned, which include known carcinogens like benzo-a-pyrene. The industry position was that the coal tar pitch volatiles (CTPVs) were responsible for the cancer excess seen in these workers. I doubted it, because men working in coke oven environments had much higher exposures to CTPVs and no excess of lymphoma or leukemia. They did, however, have a lung cancer excess.

I was suspicious that the strong magnetic fields in the potrooms might be a contributing factor. Dr. Hughes had his engineers characterize the EMF fields that the workers were exposed to.

I wanted the workers to know what we had found. There were three work shifts at the plant, so instead of making three personal presentations, Dr. Hughes and I made a video of our findings and recommendations. It was shown at the Mead plant, at the Tacoma plant, and at another Kaiser plant in Chalmette, Louisiana. A couple of days after the video was shown, I received a phone call from Don Kropp, the health and safety officer of the steelworkers union, which covered the Tacoma plant workers. "Hey doc, we have a lot of lymphoma here," he said. He was right. They had five cases of B-cell lymphoma diagnosed in a seven-year period at the plant. Dr. Hughes flew up from Oakland, and we took histories and examined the men. We also verified the pathology with referee pathologists at the University of Wisconsin medical school.

Dr. Hughes asked me to represent Kaiser in the Tripartite multi-plant study of mortality in the aluminum industry. It was called that because the aluminum companies, the steelworkers union, and the government were represented. They paid my way to meetings, but as a state employee, I couldn't take any money for my services. I was able to meet all the aluminum company medical directors and their epidemiologists. I also attended a party at the Hughes' home in Oakland, California. I was surprised that Sir Richard Doll, the English researcher, and his wife, Lady Joan, were there.

I later learned that the management of Kaiser had retained Doll to neutralize my cancer findings. I had previously spent time with Doll at Oxford University while on a lecture tour of UK medical schools. (I had an International Agency for Cancer Research fellowship and a United States technology transfer award). I was disturbed that he didn't seem to appreciate the epidemiologic strengths of my Washington State occupational mortality system. I was also puzzled by work that he and co-authors presented at a Cold Spring Harbor conference in 1981, stating that occupational exposures caused only 4 percent of all cancers and environmental exposures caused only 2 percent. I presented hard data at the same conference showing that the occupational component caused more than 10 percent of cancers.

Though confusing at the time, some of Sir Richard Doll's seemingly pro-industry stance made sense years later. After he died, it was revealed that Monsanto had paid him $1,500 per day for many years.

LEUKEMIA IN ELECTRICAL WORKERS

In 1979, Nancy Wertheimer and Ed Leeper published a study suggesting that residential magnetic fields could cause childhood cancer. They didn't actually measure magnetic fields, but noticed that the electrical distribution wires at residences with childhood cancer cases were thicker than at control houses. Thick wires carry more current, and magnetic fields are proportional to current flow.

At that time, the Kaiser engineers had measured 100 gauss (G) of static (steady) DC magnetic fields, and more than 100 milligauss (mG) of alternating AC magnetic fields in the potrooms (1,000 milligauss = 1 gauss). Most homes measure below 4 mG. Kaiser Tacoma operated on 60,000 amperes (amps) of alternating current (AC) power that was changed into direct current (DC) to supply the anodes and cathodes. I thought that if magnetic fields could cause cancer in children, what about adults working in strong fields?

When my first state occupational mortality tables became available, I looked at cancers in men who worked in jobs with a connection to electricity. These included electricians, power and telephone linesmen, aluminum workers, radio and TV repairmen, welders, power station workers, and so on. These workers turned out to have increased mortality due to leukemia, especially acute leukemia, lymphoma, and brain tumors. When the letter I wrote on these findings was published in the *New England Journal of Medicine* (Milham 1982), I received a number of calls from colleagues around the country saying that there was no way these weak fields could cause cancer. My answer was simply, "Prove me wrong." But unfortunately, over the next few months, they and others proved me right. The last time I looked, there were over fifty residential and one hundred occupational studies that now associate power frequency magnetic field exposures with cancer.

Leukemia in Amateur Radio Operators

After my letter was published, I received a note from an amateur radio operator named Andrew Sobel, W2EVE (each amateur radio operator is assigned a unique call sign). He suggested that I study mortality in members of the American Radio Relay League (ARRL). AARL is a group of amateur "ham" radio operators who are exposed to electromagnetic fields, including radio frequency radiation (RFR).

Recent ARRL deaths are reported in the "Silent Keys" section of *QST,* ARRL's monthly magazine. A quick study of the deaths reported in the Silent Keys column who were residents of Washington State and California, where I could get quick access to the death records, showed twice as many leukemia deaths as expected.

Though I had been invited by an AARL member to investigate, I did not receive the support of the organization. I negotiated unsuccessfully with ARRL for nearly a year to get access to their membership files so I could do a national population-based study. In the course of learning about the ham radio hobby, I learned that all amateur radio operators in the United States are federally licensed. I purchased the 1984 Federal Communications Commission (FCC) license file and did a cohort study of 67,829 Washington State and California amateur radio licensees (Milham 1988). Again, I chose those two states because I had ready access to death records. ARRL members did have increased mortality due to lymphoma and acute myeloid leukemia.

After the study had been published, I received a call from the National Cancer Institute (NCI) asking why I hadn't done a license class analysis too, since the exposures would be different depending on the frequency and intensity of a licensee's use. The reason I hadn't included that was because the data sheet I had received from the FCC had an undecipherable entry, " ... ass," which turned out to stand for "license class." Interestingly, when I did a subsequent license class analysis, it showed very little increased mortality in the novice entry level license class, the youngest of the five license classes, who were limited in transmitter power and restricted to certain transmission frequencies. But with increasing International Morse Code speed and theoretical knowledge, they could graduate to higher license classes and higher exposures. The higher class licensees had most of the cancer excess in the cohort (Milham 1988).

KAISER TACOMA IMMUNE STATUS STUDY

Over a ten-year period, the Centers for Disease Control assigned a series of physicians to me for a two-year training period. One of them, Dr. Robert Davis, helped me study the immune status of workers at the

Kaiser Tacoma plant in work related to the initial studies we had done there.

The steelworkers union reported additional lymphoma cases had been diagnosed since Dr. Hughes and I had examined the original group of cases. B-cell lymphoma is more common in both congenital and acquired immunodeficiency. Children born with combined immune deficiency syndrome, people with AIDS, and people treated with immunosuppressive drugs for organ transplants all have an increased incidence of B-cell lymphoma.

Dr. Hughes at Kaiser Tacoma had by then retired, and unfortunately the "risk managers" who succeeded him at Kaiser refused to cooperate with additional studies. We wanted to see if healthy aluminum reduction plant workers had any evidence of immune system problems that could account for the excess of B-cell lymphoma in the worker population. We had the support of the workers, so the steelworkers union rented a senior citizen's center near the plant. We were able draw blood samples to study the blood lymphocytes of twenty-three volunteers from the Kaiser Tacoma workforce. The University of Washington's transplant lab did a battery of blood studies on fresh blood drawn two samples at a time, which was delivered to them once a week over a twelve-week period. This meant getting up very early in the morning to meet the workers either before or after their work shift, and driving a total of one hundred twenty miles per day.

At the end of the study (Davis & Milham 1990), I received a call from the lab stating that the Tacoma workers we were studying had very abnormal T4 and T8 counts. The laboratory recommended redrawing the blood samples. With no more funds and no desire to repeat my early morning, one hundred twenty mile per day round trips, I initially refused. Later, I agreed to repeat the study if all the blood work could be drawn at one sitting with no further expense.

Dr. Davis and I took blood samples from the group again and got identical results. The T8 and T4 counts in these workers were very high, and they had abnormal T4/T8 ratios because of the high T8 counts.

My interpretation of these results was that the body was mounting an immune response to the EMF fields in the potrooms. I postulated that after time from such chronic immune system stimulation, the immune system fatigues and fails, allowing the lymphoma to occur. A later study of rats exposed to pulsed radio frequency radiation (microwaves) showed a very similar immune system pattern, and the rats also had a large cancer excess. That study is covered in detail in chapter 7.

MULTIPLE SCLEROSIS AND VITAMIN D

Early in 1980, I worked on another interesting study while at the health department. My occupational mortality data clearly showed that multiple sclerosis (MS), a remittent demyelinating central nervous system disease, was more common in indoor workers than in outdoor workers. To make sure that it wasn't a matter of self-selection into sedentary jobs, I looked at indoor and outdoor jobs in men dying of chronic rheumatic heart disease. They showed no indoor vs. outdoor mortality difference. The epidemiology of MS was fascinating. The disease was uncommon at the equator, but increased in incidence with increasing north and south latitude. Average annual hours of sunshine had a nearly perfect negative 90 percent correlation with MS incidence and mortality. This means that the more sunshine where you live, the less MS there is. The geographic exceptions were also interesting. Populations in dark regions with high fish consumption had low MS rates.

Vitamin D is really a powerful hormone, and with the exception of fatty fish and fish oils, is almost absent from the diet. Sunlight or ultraviolet light falling on the skin generates vitamin D in our bodies. Ultraviolet light is a part of the EMF spectrum that our species evolved with and obviously needs. One hour in tropical sun generates about 10,000 international units (IU) of vitamin D. In spite of this, the recommended daily dose of vitamin D is about 400 IUs. To compound the problem, sunscreens have been recommended to prevent skin cancers, cutting down on vitamin D production when outdoors. In dark climates, 3,800 IUs is a reasonable daily dose of vitamin D.

I wrote a two-page hypothesis suggesting that MS, like rickets, was a vitamin D deficiency disease. *The Lancet* rejected the paper at the time, but in the last five years, Scandinavian studies have verified all my theories about the relationship of MS to vitamin D.

At about that time my first cousin, then in his early twenties, developed MS. I explained my theory to him and suggested that he take an outdoor job. He did, and he's had minimal disease since then.

RADIAL TIRE FIELDS

My interest in electromagnetic fields grew after the Kaiser Tacoma work, and I'd followed EMF epidemiology after Wertheimer and Leeper published their 1979 study on residential magnetic field exposures and childhood cancer. I bought my first magnetic field meter in the 1990s and started attending EMF meetings to keep up on what was going on in the field. Noting that secretaries and typists had a breast cancer increase in my occupational data, and remembering the old electric typewriters used in the pre-computer era, I found that many of them had very high magnetic fields at the chest level of the typist (Milham & Ossiander 2007). The one exception was the popular IBM Selectric, which had low fields.

One day, I made the mistake of turning my magnetic field meter on in the car while driving. To my surprise, the fields were especially high (more than 20 mG) on the seat between my legs and highest near my left foot. The fields were only present when the car was in motion. Turning the engine off at sixty mph did not change the fields. It gradually became apparent that the fields were coming not from the engine, but from the spinning tires, which have a mile of fine magnetic steel wire woven into belts. I later ran a compass over the tread of a tire and watched the compass needle spin because of the magnetic domains in the belts. Electromagnetic fields can be generated by spinning magnets, which is what steel-belted radial rotating tires are.

To study the magnetic structure of the steel belts, I cut up a tire, clamped a segment down flat, covered the inside of it with white paper, and "visualized" the fields by sprinkling iron filings on the paper, which would demonstrate the magnetic pattern created by the metal belts.

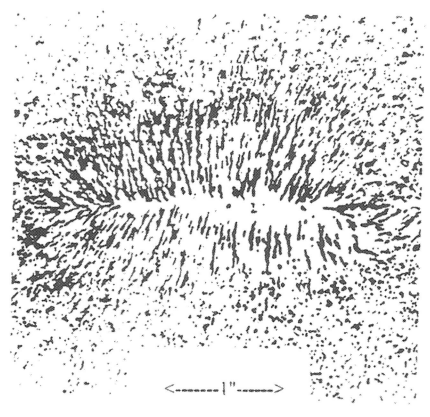

FIGURE 1
Radial tire iron filings

An X-ray of a tire segment was beautiful, showing the weave pattern of the steel belts. The radiologist smiled when I asked him to take an X-ray of a piece of tire.

FIGURE 2
Radial tire x-ray

A trip to a tire store with a compass and my inexpensive magnetic field gauss meter, along with a more expensive Emdex meter borrowed from the University of Washington, answered a lot of questions. High-end gauss meters like the Emdex have on-board computers, which allow customized long-term magnetic field sampling.

Most tires by then were steel-belted radials. My compass picked out the one fiberglass belted tire still in the shop. A spinning steel-belted radial tire on a tire balancing machine gave high magnetic field readings on my meter about a foot from the tire, but amazingly, nothing was shown on the Emdex meter. The tire frequencies are determined by the frequency of rotation, and are well below the narrow band of frequencies that the expensive Emdex meter was designed to detect. I recall that at sixty miles an hour, my Honda's tires were rotating at ten cycles a second and generating 10 Hz magnetic fields. My cheaper meter was clearly capable of picking up a wider range of frequencies.

This is of special interest, since similar high-end meters were and continue to be used in most of the human health EMF epidemiologic studies. Since these meters routinely reject all exposures below and above the power frequencies band, these studies could be missing critical exposure data. A child sitting in a baby seat over a rear tire could get a higher magnetic field exposure in an hour riding in a car than in twenty-four hours at home. The expensive meters, designed to examine a narrow band of frequencies around 60 Hz, will routinely miss fields below 30 Hz and above 300 Hz.

Eventually, I enlisted the help of electrical engineers Richard Tell and James Hatfield, who had the electrical knowledge and equipment to characterize the radial tire fields in detail. We published a paper about it in *Bioelectromagnetics* (Milham et al 1999).

I wrote to the tire manufacturers, suggesting that they could solve the problem by using non-magnetic steel wire in the belts, but got no response. I contacted the "Car Talk" radio program on National Public Radio and wrote to *Car and Driver* magazine, also with no response.

LEAVING THE HEALTH DEPARTMENT

Sometime in the mid-1980s I was informed that the health department intended to eliminate my occupational coder position. Eric Ossiander, an accomplished epidemiologist and statistician in my office, helped me automate the system. We ended up with a system that precisely matched the keyed death record occupation with those on a list. The system improved over time by adding occupational titles. It is still being used to code the decedent's occupation on the death certificate, and parental occupations on the birth certificate. It codes correctly more than 90 percent of the time. Eric Ossiander does the back up coding.

Near the end of my time at the Washington State Health Department, there was a downsizing of employees and I was demoted. I made a futile appeal to the state personnel board. In the last two years there, my office had been moved four times. I decided that since the handwriting was on the wall, it was time to leave. I resigned on the day I turned sixty, in 1992.

ON MY OWN

7

AFTER I RETIRED, I was often approached to be an expert witness, mostly testifying for people trying to prevent a power line from being sited near their homes. Like so many others who had experienced it, I did not like the legal adversarial system, or the fact that the deck was always stacked against the "little guys" who were trying to defend their personal worlds.

The utility companies had built-in rate-payer money at their disposal and hired the best attorneys, as well as very well coached expert witnesses to present their side of the case. Many times I worked for no money, only airfare and a place to stay. Unfortunately, the plaintiff's lawyers, often just local attorneys, were no match for the high-paid legal specialists from big outside firms who did nothing but EMF cases.

In one case, I was hired by the Pennsylvania Public Utility Commission to evaluate citizens' complaints about a number of proposed high voltage transmission lines. This was the only case that I was involved in where taxpayer money was actually used to support and evaluate taxpayers' concerns. I helped stop a few power line corridors from being built temporarily, but knew that the utilities would ultimately prevail, as they typically did.

I saw the same handful of utility "experts" in all of these cases. I often wondered how they could look themselves in the mirror after making misleading comments under oath, saying that power line EMFs were safe. It was also galling that they were so well paid with rate-payer money to testify against the rate-payers' best interest.

At breakfast one morning during one of the trials, a utility expert I'd known for twenty years admitted that our side was right. But it did not stop him from pocketing a reported quarter of a million dollars a year as an expert witness for the utility companies, in addition to his full university professor's salary.

MALE BREAST CANCER AND EMF EXPOSURE

I testified in a trial for two of three men who had developed breast cancer from sitting in the same office in a county office building in Albuquerque New Mexico, where they were exposed to strong magnetic fields from a nearby electrical switchgear room (Milham 2004). A dozen previous studies had linked male breast cancers (MBC) to EMF exposures, so these cases should have been a slam dunk. The plaintiff's attorney had already won a nice settlement for one of the three men, but didn't seem to care about the other two cases. He didn't cross-examine the defendant's EMF expert, and cut my testimony short to celebrate a religious holiday. Without a good attorney, an expert witness is both helpless and useless.

There are reports of an epidemic of male breast cancer at the Marine Corps Base Camp Lejeune in North Carolina in the environmental news right now. The Marines had identified fifty-five MBC cases and thought the epidemic was caused by solvent contamination of drinking water at the base. While there are studies that link solvents with a few cancers, the more likely culprit is EMFs. There have been fifteen studies linking MBC with EMF exposure. This is so unusual and so consistent that I'd consider male breast cancer a sentinel cancer for EMF exposure like mesothelioma is a sentinel for asbestos exposure. Since male and female breast cancer probably share causal relationships, study of the very rare male breast cancers may yield important clues about the etiology of the much more common female breast cancer. My 2010 letters to the Marines at their Web site, to their attorneys, and to the commandant at Camp Lejeune have not been answered.

I am sorry to report that President Obama has now signed into law the "Honoring America's Veterans and Caring for Camp Lejeune Families Act of 2012" which directs the Department of Veterans Affairs to provide VA healthcare to veterans and their families who have been diagnosed with a disease related to the toxic water contamination that occurred at Marine Corps Base Camp Lejeune from 1957 to 1987.

CANCER IN FIREFIGHTERS

Recent reviews and reports of cancer incidence and mortality in firefighters conclude that they are at an increased risk of a number of cancers. These include leukemia, multiple myeloma, non-Hodgkin's lymphoma, male breast cancer, malignant melanoma, and cancers of the brain, stomach, colon, rectum, prostate, urinary tract, testes, and thyroid. Firefighters are exposed to a long list of recognized or probable carcinogens in combustion products and flame retardants and the presumed route of exposure to these carcinogens is by inhalation. Curiously, respiratory system cancers and diseases are usually not increased in firefighters the way they are in workers exposed to known inhaled carcinogens like roofers, smelter workers, and asbestos and insulation workers.

The list of cancers showing an increased risk in firefighters strongly overlaps with the list of cancers with high rates in workers exposed to electromagnetic fields (EMF) and radiofrequency radiation (RF). Firefighters have increased exposure to RF in the course of their work, from mobile two-way-radio communication devices that they routinely use while fighting fires, and at times from firehouse and fire vehicle radio transmitters and pagers (Milham 2009).

Recent studies of long-term California firefighters working in fire stations near cell towers have shown that they have many of the signs and symptoms of microwave illness like sleep deprivation, depression, headaches, tremors, and slowed reaction time. Additionally, they had abnormal functional brain scans. The International Association of Fire Fighters has a resolution banning fire station cell towers pending further research.

I suspect that some of the increased cancer risk in firefighters is caused by both RF and dirty electricity exposure and is therefore preventable. The precautionary principle should be applied to reduce the risk of cancer in firefighters, and workmen's compensation rules will necessarily need to be modified.

FLOUR MILL WORKERS AND ALUMINUM WORKERS FATHERS

Analysis of the parental occupation data on the Washington State birth record has yielded new knowledge a number of times. It uncovered an increased incidence of female births to fathers whose occupation was recorded as "aluminum worker" or "flour mill worker." The aluminum workers have strong EMF and heat exposures, both of which are known to damage sperm. Flour mill workers are exposed to pesticides used to prevent insect infestation in flour. Dichlorodibromopropane (DCBP), known to cause infertility and sterility in men exposed to it, is the probable cause of the sex ratio change in children of flour mill worker-fathers. Their sons also have significantly lower birth weight (Milham 2009). There may be parallels between EMF exposures and toxic exposures.

CANCER IN OFFICE WORKERS

In the mid-1990s, I received a call from a California attorney who was representing workers at a Grubb and Ellis commercial real estate office. The workers there claimed there were too many cancer cases among them. Newly installed computers had vibrating screen images, which occur in high magnetic field environments.

They worked on the first floor of a fourteen floor office building, and sat over three twelve-kilovolt (kV) transformers that caused strong magnetic fields in their office space. The employer, the building owner, and the utility company had hired experts to conduct a cohort study of cancer incidence in the workers, and they had reported that there was no excess of cancers beyond what would be statistically expected.

Curious about their findings, I agreed to look at the data. About thirty seconds after loading their CD, I noticed that most of the cancer cases were in long-term workers, in spite of the fact that more than half of the workers had worked there for less than two years. An analysis of duration of employment showed a positive linear relationship between duration of employment and cancer risk, meaning the longer someone was employed there, the more likely they were to develop cancer. If a risk of one was set for workers employed for less than two years, the workers employed for two to five years had a cancer risk of 9.3; and workers employed from five to fifteen years had a cancer risk of 15.1. Risks this high are very unusual in most power-frequency magnetic field cancer studies.

Two other findings in this group were unusual. The all-cancer incidence for both sexes was elevated, and malignant melanoma was over-represented. There were two melanomas among the eight invasive cancers, and two additional melanomas that were limited to the outer skin layers (in-situ) that were not counted in the cohort analysis. I contacted the people who had done the study and told them what I had found. Since it was their data and analysis, and I felt odd reporting on their data. They didn't want any attribution (or blame), so I published my analysis without revealing who they were, and didn't identify the workers or the building (Milham 1996).

UNIVERSITY OF WASHINGTON AND MICROWAVE-EXPOSED RATS

Though not well known, there is a group of academic, military, and industry researchers who have been paying attention to this subject for decades. It's called The Bioelectromagnetics Society (BEMS), and among other activities, they publish a peer-reviewed journal. I attended a few BEMS meetings, but eventually resigned when industry influence began to heavily permeate their journal. I wasn't alone in leaving over this problem. Many are also unaware that the Russians and Eastern Europeans have been at the forefront of this research for many years, and have enacted stricter EMF exposure standards than in the west.

At one memorable BEMS meeting in the late 1970s, I was seated next to Louis Slesin, publisher of *Microwave News*, the best newsletter for keeping up on all aspects of EMF research. A University of Washington research group, headed by William Arthur Guy, presented the results of a $4.5 million study that they had conducted for the United States Air Force, using a rats in a germ-free exposure environment to reduce environmental variability. One of their slides showed eighteen cancers in the one hundred exposed rats, but only five cancers in the one hundred sham-exposed controls. A quick calculation told me that this difference was very unlikely to have happened by chance. Nevertheless, the authors were declaring this a "negative" study.

The exposures in the experiment were to non-thermal (non-heating) levels of pulsed and modulated RF at microwave frequencies (2450 MHz). I asked a question during the meeting about it, and Lou Slesin and I met with the authors after the presentation. The authors danced around this striking finding with completely illogical arguments. First, they claimed that the control cancer level was too low, and then that we couldn't add all the cancers together, since they were of different types. It turns out that most of the cancers were of endocrine system organs.

This was the first, and to date only, long-term, low-level animal study of non-thermal microwave exposures done in the United States. In their scientific publications at the time, and still in their interpretation today, this group refuses to acknowledge that this well-done study showed RF microwave exposure to be a potent animal carcinogen. In addition, at midlife, their exposed rats also had significant elevations of immune system T and B cells very similar to the Kaiser Tacoma aluminum workers. Like the aluminum workers, the rats were mounting an immune response to the microwaves. I think that with chronic immune stimulation, the immune system fatigues and fails, allowing the cancers to occur.

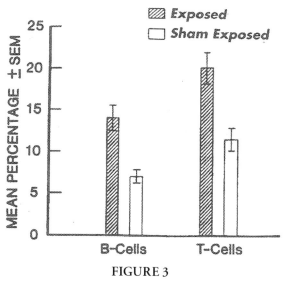

FIGURE 3

T and B cell counts in microwave exposed and sham exposed rats

There also were seven benign adrenal pheochromocytomas in the exposed rats, versus one in the controls. These are functional tumors of the adrenal medulla, which produce adrenaline. Eric Ossiander, my friend from the health department, provided me with the annual number of pheochromocytoma hospital discharges in Washington State between 1987 and 2007.

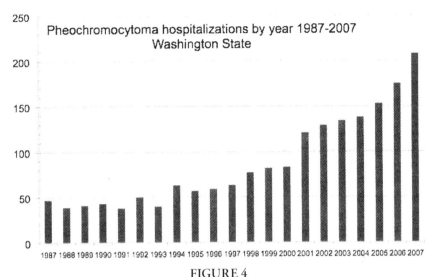

FIGURE 4

Pheochromocytoma hospitalizations WA

There has been a remarkable increase in cases, with a steep increase in annual cases beginning in the year 2000, just as cell phones were becoming most popular. For example, there were eighty-three hospital discharges in Washington State for this tumor in 2000, but 208 in 2007. National United States hospital discharges show a similar pattern with 1,927 total discharges for benign adrenal gland tumors in 1997, and 3,764 in 2007. Clearly something is going on, which may be a reflection of rapidly increasing ambient RF levels in our environment.

CHILDHOOD LEUKEMIA AGE PEAK STUDY

One night in 1999, while surfing the Web, I came across a graph (Figure 8, chapter 9) showing the percent of urban and rural residences connected to the electric utility grid in the mid-1900s (Ausubel and Marchetti 1996). I E-mailed the author, Jesse Ausubel of Rockefeller University, and he gave me a reference to a United States Census Bureau document. The Michael Court-Brown and Richard Doll paper documenting the emergence of the U.S. childhood leukemia age peak in the 1930s (Court-Brown and Doll 1961) came to mind, and I intuitively thought that electrification might have caused the emergence of this childhood peak.

The childhood leukemia age peak simply didn't exist in places without electricity, and electrification is worldwide phenomenon. The only other technology to cover the planet in the last century was the internal combustion engine, and it was present in rural areas of the United States long before electricity. Some change in the environment had to account for this steady rise in childhood leukemia incidence.

Eric Ossiander and I dug out childhood leukemia deaths for children dying under age five classified by state in the years around the 1930 and 1950 U.S. censuses. From the United States Census Reports, we abstracted individual state population data and the percentage of homes with electricity in urban, rural non-farm, and farm homes. At ages two through four in the peak years, there was a 24 percent increase in leukemia mortality for every 10 percent increase in the percentage of homes served with electricity. There was no relationship to electricity in the non-peak ages from birth through age one.

The peak was made up of a single leukemia subtype, common acute lymphoblastic leukemia, and the rates worldwide vary from 0.4 per 100,000 people in places without electricity, to more than 4 per 100,000 people in industrialized countries.

Trying to get the paper published was an ordeal. We tried the *American Journal of Epidemiology* (AJE), the *Journal of the American Medical Association (JAMA),* the *Lancet,* and others. I sent it to a friend who had been editor of AJE. She commented that there was nothing wrong with the paper, but that it might have been too "politically challenging." She suggested sending it to *Medical Hypotheses,* where it was finally published (Milham & Ossiander 2001).

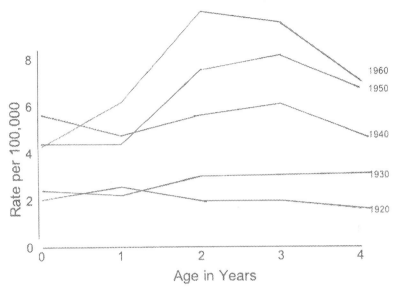

FIGURE 5
Childhood leukemia mortality rate by single years
of age for census years 1920–1960

The electrification status of individual farms had been collected in the 1930 census farm schedules. We thought that calculating childhood leukemia rates for electrified versus non-electrified farms would really nail the association. The Census Bureau, however, delayed release of some data for seventy years, so in 2000 I asked them for the 1930 farm schedules. I was informed that they had been destroyed or lost or both.

I then searched the Web for another place that might have a detailed electrification history, and found that a provincial utility company in the Canadian province of Manitoba had a record for the year when 523 towns in the province had been hooked into the utility grid between 1920 and 1961. I jumped through all of the privacy and research design hoops, but at the last minute I was denied access to the provincial vital records data. I would predict that if anyone ever takes a look, childhood leukemia deaths will likely follow the northward spread of electrification from Winnipeg to Churchill, Manitoba.

While doing the childhood leukemia study, I also noticed a strong positive correlation between residential electrification and mortality for with some adult cancers, including female breast cancer, in the 1930 and 1940 vital statistics. At the time, there was no single plausible human electrical exposure agent, or a method for its delivery. It just didn't seem possible that power-frequency magnetic fields alone could be responsible for these increased deaths, even though the association between mortality and residential electrification was clear.

The answer would come in 2008 with my study of a cancer cluster in teachers at La Quinta Middle School in La Quinta, California, covered in the next chapter. That study would point to high frequency voltage transients, also known as "dirty electricity" as a potent, universal carcinogen.

CANCER AT THE LA QUINTA MIDDLE SCHOOL

In **FEBRUARY 2004,** an article in the Palm Springs, California, newspaper *The Desert Sun* came to my attention. The headline read "Specialist discounts cancer cluster at school." It was written by Michael Perrault. The specialist was John Morgan, a Ph.D. epidemiologist who directed the Desert Sierra Tumor Registry at Loma Linda University Cancer Research Institute, and the school was the La Quinta Middle School (LQMS) in La Quinta, California, located about four miles west of where we live in winter.

The article by Perrault listed eleven cancers in the teachers by type, and listed the names of three teachers who had died of cancer. Dr. Morgan had come to the school at the request of the school district and talked to the teachers to allay their fears. Later, I saw a video of that meeting and counted twenty or so times in which he told teachers that they had no cancer cluster, or that the incidence of cancer at the school was "normal."

Gayle Cohen, a teacher then being treated for breast cancer, was the main representative for the teachers. Perrault's article defined what a cancer cluster is, but two things raised my suspicion. There was no mention of how large the school was, or how many teachers taught there. These figures are critical in analyzing cancer clusters. There was also no mention that Morgan had done any sort of study.

Five minutes of computer searching really got me interested. The school opened in 1988, was incorporated into a new building in 1990, and had thirty-seven teachers in 2004. A quick mental calculation told me that the teachers were right. There were a lot more cancers in their group than would be expected, based on the size of their staff, a guess about their ages, and knowledge of cancer incidence rates.

I called Mr. Perrault at *The Desert Sun,* told him that I thought the teachers had a cancer excess, and left a phone number for the teachers to call if they wanted help in testing their suspicions. Gayle Cohen called and said that the teachers would welcome my help. I was disturbed when they reported that Dr. Morgan had not contacted a single teacher in the group personally, and had seemingly rendered his "no-cluster, no-cancer-increase" message without any evidence. A number of the La Quinta teachers had taught at other schools and continued to know teachers at other schools. In no other school were they aware of more than one or two cancer cases. At La Quinta, they knew of eleven cancer cases.

To do a proper study of whether or not there was a cancer excess at the school is a fairly straightforward exercise. We needed a list of all the teachers who had ever taught at the school, their birth dates or ages, hire and termination years, vital status, and cancer status. Cancer diagnoses had to be verified with pathology reports, and reconciled with the regional tumor registry. Population cancer rates, categorized by age, race, and sex, were available online and could be used to calculate expected cancer cases, based on the age, sex, and duration of employment of the teacher population.

I called the school and was referred to Charlene Whitlinger, Assistant Superintendent of Schools for the Desert Sands Unified School District. When I reached her by phone, I was curtly told to put my request in writing. On March 11, 2004, I sent her a copy of my curriculum vitae and a one-page proposal describing my estimates of cancer excess, which clearly differed from what Dr. Morgan had told them, and asked to make magnetic field measurements at the school. After a number of futile phone calls over the next few weeks, a kind person at the school informed me that my proposal would never be answered. I then E-mailed the superintendent of the school district, Dr. Doris Wilson, and got a quick response on April 22, saying, "We feel that at this time our investigation and findings are satisfactory."

Over the summer, the teachers, especially Gayle Cohen and the widow of one of the La Quinta teachers, Linda Loveless, herself a teacher at another school, developed a teachers' roster using school yearbooks.

Also over that summer, I had chatted with Lloyd Morgan, a retired electronics engineer who I had met at BEMS meetings. Lloyd had recovered from a large frontal meningioma (benign brain tumor) that he thought was caused by sleeping for years with his head just inches from a bedside electric clock. Lloyd had been working with Professor Martin Graham and Dave Stetzer, co-inventors of a meter that measures high frequency voltage transients ("dirty electricity") when plugged into electrical outlets. They also manufactured a plug-in filter that could reduce the dirty electricity levels.

The meter invented by Graham and Stetzer displays the average rate of change (dV/dT) of these high frequency voltage transients that exist everywhere on electric power wiring today. High frequency voltage transients found on electrical wiring, both inside and outside of buildings, are caused by an interruption of electrical current flow, essentially creating high-frequency voltage transients that should not be present on normal 60-Hz sine waves used in power distribution.

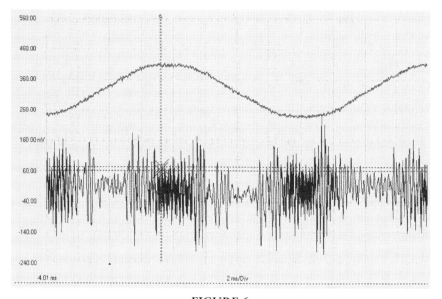

FIGURE 6

*Oscilloscope tracing: channel 1 is 60-Hz AC line voltage; channel 2
is channel 1 X 10 with the 60-Hz removed with a high pass filter*

The upper line in Figure 6 is the sixty cycle sine wave of AC power. The bottom line displays the high frequency voltage transients enlarged ten times, which are riding along on the sine wave.

There are many sources of dirty electricity in today's electrical equipment. Examples of electrical equipment specifically designed to operate by interrupting current flow include: light dimmer switches that interrupt the current twice per 60-Hz cycle (120 times per second); power-saving compact fluorescent lights that interrupt the current at least 20,000 times per second; halogen lamps; electronic transformers; and most electronic equipment manufactured since the mid-1980s that use switching or switch mode power supplies. Older fluorescent lighting systems also generate dirty electricity.

Except for the dimmer switches, most of these devices didn't exist in the first half of the twentieth century. However, early electric generating equipment and electric motors used commutators, carbon and metal brushes, and split rings, which would inject high frequency voltage transients into both the original DC current and then the AC 60-Hz electricity that followed. Modern electronics virtually guarantee creation of high levels of dirty electricity, since most electronic devices use switching power supplies. The recent utility practice of tying the neutral return lines to the ground will also increase dirty electricity in ground currents.

A 1994 study of Canadian and French electric utility workers furnished a tantalizing clue about the carcinogenicity of high frequency transients. A study sponsored by Hydro Quebec, a Canadian power utility, showed a fifteen-fold increased risk of lung cancer in workers exposed to pulsed high-frequency magnetic fields, with a rising incidence according to dose. Risk ratios this high are almost never seen in studies using power-frequency magnetic fields as a metric. These findings were independent of a worker's smoking status. Unfortunately, Hydro Quebec sequestered the data and disappeared the Positron meter used to make the exposure measurements (Armstrong, et al. 1994), so no replication or follow-up was possible by anyone else.

Dirty electricity generated by electrical equipment in a building is distributed throughout the entire building via the electric wiring. Dirty electricity generated outside the building can enter a building on electric wiring, or through ground rods and conductive plumbing. Over the last twenty years, about 70 percent of electricity returns to the substation via the earth rather than through the neutral transmission lines. In remote, sparsely populated areas, single wire earth return systems return all current through the earth. When the grid was built, it was designed for all current to return in the neutral wires. The grid just couldn't keep up with increasing return current loads, so the utilities were allowed to begin using the earth for return currents. Every other power pole in most areas has a wire running down it connecting the neutral to the ground.

Within buildings, dirty electricity is usually the result of current flow being interrupted by our own electrical appliances and equipment.

Arcing, sparking, and bad connections in wiring can also generate dirty electricity.

Human exposure occurs by capacitive coupling of the ambient high frequency fields with the body. These fields generate exogenous currents in the body.

INVESTIGATING THE LA QUINTA MIDDLE SCHOOL

Since I'd already published the paper on the cancer cluster in the real estate office where the workers were exposed to strong magnetic fields from the three 12 kV transformers directly below their work area, I was anxious to measure the magnetic fields in the La Quinta Middle School (LQMS). Lloyd Morgan agreed to accompany me to visit the school, at the invitation of Gayle Cohen, to make magnetic field and dirty electricity measurements.

Lloyd Morgan had the Graham/Stetzer meter and an oscilloscope with which to visualize the electricity flowing on the school wiring. We both had magnetic field meters. In February 2005, over a two-day period, Lloyd and I arrived at the front desk at LQMS after school hours, signed in, and had Gayle Cohen paged. Gayle showed us the way to her classroom and had a janitor open doors for us.

We found very high power-frequency magnetic fields in room 304, which were caused by proximity to the main electric service room for the entire school. But curiously, magnetic fields in the classroom space in most of the seven rooms we surveyed were normal. What we did find was what we thought could be a net current problem caused by unbalanced currents in a couple of other rooms. However, the main finding with the Graham/Stetzer (G/S) meter and the oscilloscope was very high levels of dirty electricity in the outlets of all of the rooms we visited. Every room had high levels of dirty electricity.

While most homes or businesses typically read under 100 G/S (Graham/Stetzer) units, the school buildings averaged about 700 G/S units. Ideally, readings should be under 50. The version of the meter we used gave an overload indication at 2,000 G/S units, and many of the outlets were overloaded. Since the net current problem is a potential fire hazard, I thought it wise to let Doris Wilson, the superintendent of the school district, know what we found and wrote her on February 27, 2005, with our findings.

On March 10, 2005, I received a certified letter from her, obviously written by an attorney for the school district, using terms that claimed my visit to the school was, "A clear violation of ... state law, unlawfully trespassed ... dangerous and destructive testing ... safety of students ...." It even cited Homeland Security concerns. Gayle Cohen received a letter of reprimand. The school district hired an independent electrical contracting firm and they came up with exactly the same magnetic field results that we had reported, but they did no readings for dirty electricity.

I contacted Dr. Raymond Neutra of the California Department of Health Services (DHS) in March 2005. I had known him for more than twenty-five years and knew that he had reviewed the general EMF literature over the years as part of a state-mandated project. He agreed that my back-of-envelope calculations showing a cancer excess at the school was correct, but indicated that the regional tumor registry would have to certify that the cancer excess existed before he could get involved.

I also had a personal meeting that spring with the president of the local teachers' union and a representative of the California State Teachers Association. In March 2005, I had the teachers file a federal complaint with the National Institute of Occupational Safety and Health (NIOSH), which was ignored, but in May 2005, I had the teachers file a state complaint with the California Occupational Safety and Health Administration (CAL OSHA), which got Dr. Neutra and the California DHS involved. Every active teacher at the school signed the complaint.

On May 16, 2006, Dr. John Morgan, the cancer registry

epidemiologist, and I gave presentations about the LQMS cancers at a meeting of the Desert Sands Unified School District School Board. Dr. Morgan (no relation to Lloyd Morgan) agreed that there was a statistically significant excess of malignant melanoma among the LQMS teachers, but claimed that there was no excess of all cancers. After the meeting I found out that he had used unpublished cancer rates to calculate expected cancers. Nearly every one of the rates he used was systematically higher than the official published rates that we had used. His estimates of expected cancers were, therefore, inflated. Additionally, he blamed the melanoma excess on the fact that the desert is a "sunny place"

On June 8, 2006, Dr. Raymond Neutra and contractors for the school district took magnetic field measurements in most of the classrooms at LQMS. Most importantly, Dr. Neutra used the G/S meter to determine the dirty electricity level in many of the outlets in most of the classrooms at the school. (Dr. Neutra knew Dr. Martin Graham and had encouraged use of the G/S meter in a large study then being conducted for the large health cooperative, Kaiser-Permanate, on spontaneous abortions.) This gave us one piece of the exposure data that was needed to create a good exposure-based study. The other important piece of exposure data was furnished by a teacher who had saved a list of every teacher's classroom assignment since 1990. This located each teacher in a room at the school for which we now had exposure measurements.

Thirteen of the fifty-one classrooms at the La Quinta school had dirty electricity levels above 2,000 units. The levels averaged about 700 units. By contrast, none of forty-one rooms at a Washington State school had a reading above 100 units.

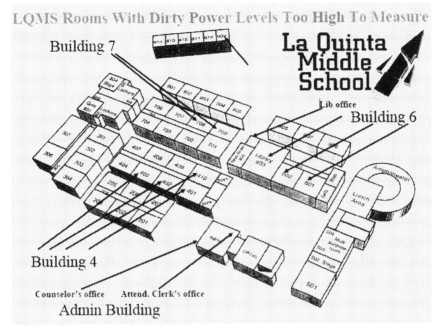

FIGURE 7
Layout of La Quinta Middle School showing classrooms measuring above 2,000 units on the Graham-Stetzer meter

I suspect that the dirty electricity levels at La Quinta might be originating from a defective utility substation about a mile away, carrying transients on the lines into the school. The levels are higher than what ordinary school electronics would create and high levels are present in other buildings in the vicinity. There was strong AM radio interference along the transmission lines all the way from the substation to the school.

Over the next six months, Dr. Neutra and a health department statistician worked on an analysis of the cancer and teacher information that Lloyd Morgan and I had provided them. They used an approach that compared the most highly exposed teachers to those with lower exposures. Since even the lowest exposed teachers had a cancer increase, this method was going to give lower risks. The method we used generated expected cancers based on California cancer incidence rates, specific for age and sex. I also sent the data to Dr. Gary Marsh at the University of Pittsburg, who ran an occupational software program called OCMAP on our data. He confirmed all our results.

OUR FINDINGS

Sixteen schoolteachers in a cohort of 137 teachers who had ever been employed from La Quinta's opening in 1988, through December 2005, had been diagnosed with eighteen cancers. (Two teachers each had two cancers). The observed-to-expected (O/E) risk ratio for all cancers was 2.78, while the O/E risk ratio for malignant melanoma was 9.8. Thyroid cancer had a risk ratio of 13.3, and uterine cancer had a risk ratio of 9.2. All of these numbers represent greatly elevated risks. Most EMF-cancer studies measuring power-frequency magnetic fields almost never find risks above 4.

What we discovered surprised us. Contrary to all of the recent research published on EMFs, the 60-Hz magnetic fields showed no association with cancer incidence. The new exposure metric, high-frequency voltage transients known as dirty electricity did show a positive correlation to cancer incidence. In addition, a cohort cancer incidence analysis of the teacher population showed a positive trend of increasing cancer risk with increasing cumulative exposure to high-frequency voltage transients on the classroom's electrical wiring as measured with a G/S meter. Cancer risk also increased with duration of employment.

The attributable risk of cancer associated with this exposure was 64 percent. Attributable risk simply answers the question of what percent of a disease is due to an exposure. This calculation subtracts the expected cancers from the observed cancers and divides that difference by the observed cancers. According to this calculation, a single year of employment at this school increased a teacher's cancer risk by 21 percent. A single year of employment in a room that had a G/S meter overload increased a teacher's cancer risk by 26 percent. These cancer risk estimates are probably low because twenty-three of the 137 members of the cohort were lost to follow-up. They were counted in the expected cancer risk calculations, but could not add cancer cases.

In addition, since exposure was calculated based on seven days per week for a twelve-month year, this would overstate the actual teachers' exposure of five days per week for nine months a year, again leading to an underestimation of cancer risk. A peer-reviewed paper reporting our findings was published in 2008 in the *American Journal of Industrial Medicine* (Milham & Morgan 2008).

THE FALLOUT

In the spring of 2007, the school district personnel, their attorneys, and State Senator James Battin traveled to Sacramento to meet with Sandra Shewry, who was Dr. Neutra's boss at the California DHS. I assume they were there to discuss his final OSHA report about cancer in the school teachers, which by then had been released.

In his report, Dr. Neutra agreed that the cancer incidence among the teachers was high, that the dirty electricity levels at the school were exceptionally high, but he waffled at actually linking the cancers to the dirty electricity levels. His words in his report were, "I would say that I am prone to doubt this hypothesis but I am not virtually certain that it is not true. Our review of the air and water information you provided us yielded no information that would have explained the cancers at the school. Nevertheless, there is an excess of cancer at La Quinta Middle School that is unlikely due to chance and is associated with circuit vibration …," his more formal term for dirty electricity.

Given their already demonstrated mindset, the school district interpreted Dr. Neutra's report as a clean bill of health for the school. At the end of April 2007, at the request of the teachers, Lloyd Morgan and I held a public meeting to give the teachers, parents, and community a final report of our study. Charlene Whitlinger, the deputy superintendent of schools, attended briefly, but was heckled out by the teachers and parents.

In January 2008, a new school district superintendent, Sharon McGehee, was hired to replace Doris Wilson, who had retired. I wrote her a friendly letter in the naïve hope that she might finally act to help the teachers at LaQuinta. She never answered me. Instead, I received a letter from Mr. David G. Miller of the Los Angeles law firm Miller, Brown, Dannis telling me to "cease and desist." He advised me not to contact the school district directly in the future, but to contact him instead. After accusing me of using the school as a "guinea pig" for my own ends, he ended his letter rather dramatically with, "We cry Enough. Enough, Dr. Milham, Enough!"

In the end, the school did spend a small fortune to shield room 304 from high magnetic fields. For about $5,000, they could have filtered the whole school and removed the dirty electricity hazard.

TYING UP LOOSE ENDS

When the total cancer rate and melanomas turned up high at LQMS, I remembered that that was the same pattern in a cancer cluster I had reported in workers at the first floor real estate office in the high rise building (Milham 2008). I Googled the Grubb and Ellis Company, who had occupied the offices at the time of my study, and found out that they had moved to new quarters. After talking to one of their employees on the phone, I received a letter from their attorney and was told that any future contact should be through her. The president of the company currently occupying the office space, Conexant Corporation, never answered my letter. People were clearly afraid of such findings, likely due to liability for their employees.

The pattern of cancer increase at LQMS is also identical to that found in a large study of cancer incidence in California Teachers

Association (CTA) members in the late 1990s (Reynolds, et al. 1999). I contacted the CTA and one of their attorneys met with me and some of the LQMS teachers in the early spring of 2008. After not hearing back from him, I called to see what the problem was, and learned that he felt that the teachers didn't have a case worth pursuing. Letters to the president of the CTA have gone unanswered.

The sad part is that the LQMS story is probably being repeated in other schools across California and throughout the world. And it's probably not just teachers who are at risk. Anyone near wires carrying dirty electricity is at risk, too.

An open question from the beginning has been whether the students who spend three years at LQMS are at risk. In April 2008, I learned of three former students in their mid-twenties who had been diagnosed with thyroid cancer, and another who had died of breast cancer. There was also a thirty-year-old woman who had two primary invasive malignant melanomas and in 2009 had both breasts removed (one prophylactically) for breast cancer in one breast. It was obvious that three years at LQMS is enough time and exposure to cause cancer both in the teachers and students.

The La Quinta study would have been impossible without the pioneering work done by Dr. Martin Graham and Dave Stetzer. Dr. Graham is an emeritus professor of electrical engineering at UC Berkeley. In the 1990s, he was an expert witness for dairy farmers in California who were suing milking-machine manufacturers for reduced milk production and health problems in dairy cows. In 1995 he patented a device for measuring and monitoring electric current flow in cows. He teamed up with Dave Stetzer, a power quality electrician from Blair, Wisconsin, who was dealing with similar problems in farm animals, as well as health problems in farm families. Their genius was to recognize that it was the high-frequency voltage transients in ground currents that were causing the problems.

Dr. Graham and David Stetzer co-authored a paper with Dr. Donald Hillman, professor emeritus at the Department of Animal

Science at Michigan State University with a very descriptive title: "Milk Production of Dairy Herd Decreased by Transient Voltage Events." Dr. Graham patented the Graham/Stetzer meter that measures transients on electric wires by plugging into an outlet. They also developed a plug-in capacitor filter for reducing transients.

Dr. Magda Havas, a professor at Trent University in Peterborough, Ontario, Canada, became involved with Dr. Graham and David Stetzer when she discovered dramatic changes in symptoms and student behavior at a Toronto School after she measured dirty electricity and deployed their filters to reduce it. Dr. Havas has done important studies on the health effects of dirty electricity and may have identified a third kind diabetes associated with high frequency transients, in addition to the recognized type 1 and type 2 diabetes (see: http://www.stetzerelectric. com/filters/research/).

If Dr. Raymond Neutra hadn't known Dr. Graham and had not used the G/S meter to measure the transients at the La Quinta Middle School, we would never have been able to uncover the cancer connection to dirty electricity. Dirty electricity has frequencies between 2 and 100 KHz in the radiofrequency range. These fields are potentially present on all wires carrying electricity and are an important component of ground currents returning to substations, especially in rural areas. This helped explain why the professional and office workers, like the school teachers, had high cancer incidence rates. It also explained why indoor workers in my Washington State studies had higher malignant melanoma rates, and why melanoma in general can occur on parts of the body that never are exposed to sunlight. It also explains why melanoma rates are increasing while the amount of sunshine reaching earth is stable, or even decreasing due to air pollution.

A number of very different types of cancer had elevated risk in the La Quinta school study, in the California school employees study, and in other teacher studies. The only other carcinogenic agent that acts like this is ionizing radiation.

THE DISEASES OF CIVILIZATION

DISEASES OF CIVILIZATION or lifestyle diseases, also sometimes called diseases of longevity, are diseases that increase in frequency as countries industrialized. They include cancer, Alzheimer's disease, cardiovascular diseases, asthma, type 2 diabetes, obesity, osteoporosis, and depression. Although known in antiquity, the incidence of these diseases has increased steadily in the twentieth century and continues to increase in the new millennium. The prevailing wisdom is that these diseases are due to changes in diet, sedentary lifestyle, increased crowding in cities, cigarette and alcohol consumption, and disintegration of societal and family support. The childhood leukemia connection to residential electrification and the La Quinta dirty electricity connection to adult cancer made me think that that the diseases of civilization might instead be diseases of electrification.

With the new exposure metric of dirty electricity and the means for its delivery via our utility lines and ground currents everywhere, I decided to examine whether residential electrification in the United

States in the first half of the last century was related to any other causes of death besides childhood leukemia.

This new study was done by keying death and population data into Excel spread sheets, calculating death rates (deaths/population x 100,000), running correlation software, and looking at the relationship between the percent of electrified residences by state and the death rates for urban and rural areas. It took approximately three months, sometimes using a magnifying glass to look at tiny, faded numbers in seventy-year-old books.

Examination of adult mortality from all causes in the 1930 and 1940 United States vital statistics provides evidence that residential electrification was responsible for the epidemic of our diseases of civilization in the twentieth century. The slow spread of residential electrification in the United States in the beginning of the twentieth century from urban to rural areas resulted in two large populations by 1930. There were urban populations with nearly complete electrification, and rural populations that were exposed to varying levels of electrification, depending on the level of electrification in their state.

Growth of residential electric service in US states from 1920 to 1956.

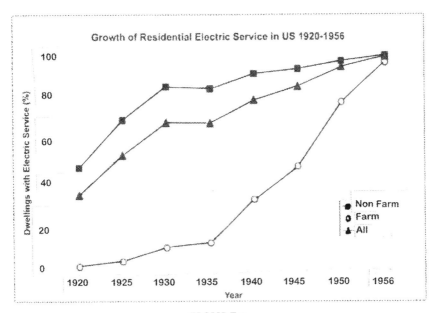

FIGURE 8

Percent of farm and non-farm dwellings with electric service 1920–1956 U.S. Census Bureau data

Both populations were covered by the United States vital registration system, so the data was good. By 1940, all of the contiguous forty-eight states were included, and deaths were counted in separate books according to place of residence and place of death. Since 1900, there has been a gradual increase in mortality rates of cancer, cardiovascular disease, diabetes, and suicide, the so-called diseases of civilization. This is in sharp contrast to the gradual decline in the death rate from all causes, which was reflecting increasing control of infectious diseases.

Figure 9.——Death Rates: Death-registration States, 1900–32, and United States, 1933–60

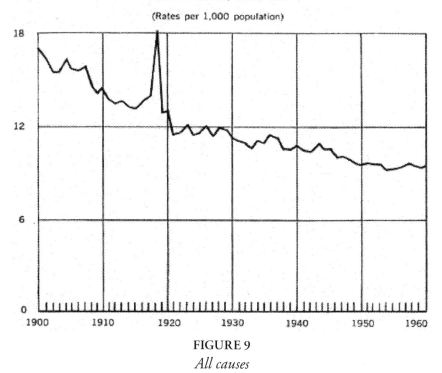

FIGURE 9

All causes

Figure 10 –Death Rates for Major Cardiovascular-renal Diseases: Death-registration States, 1900–32, and United States, 1933–60

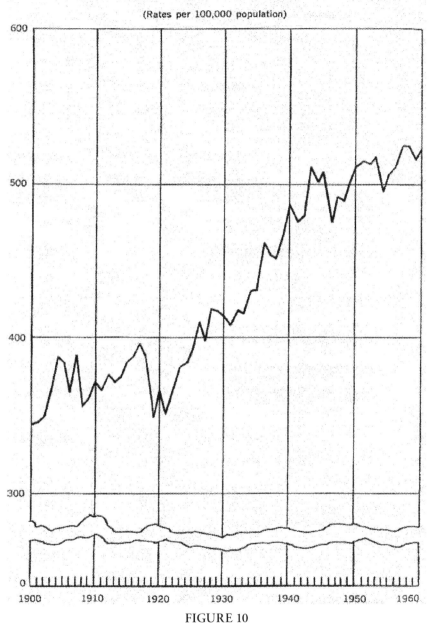

(Rates per 100,000 population)

FIGURE 10

Cardiovascular

Figure 11 –Death Rates for Malignant Neoplasms: Death-registration States, 1900–32, and United States, 1933–60

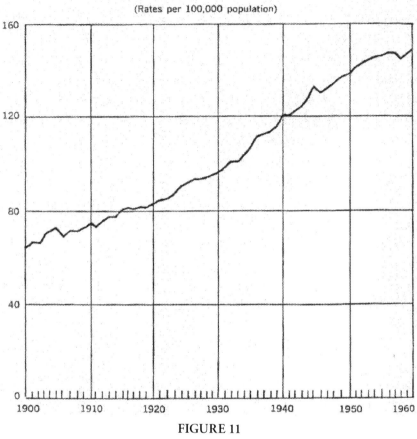

(Rates per 100,000 population)

FIGURE 11

Malignant neoplasms

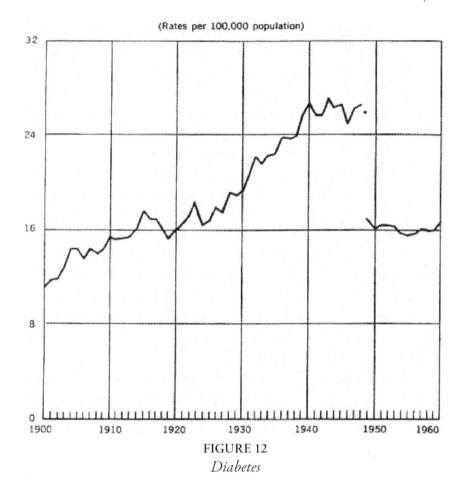

(Rates per 100,000 population)

FIGURE 12
Diabetes

Figures 9–12 were scanned from *Vital Statistics Rates 1940–1960 in the United States* by Robert Grove Ph.D. and Alice M. Henzel. This volume was published in 1968. In Figure 12, the break in the rate line is due to a code change

I abstracted, keyed, and analyzed United States vital registration mortality information and population census information for 1930 and 1940. Death rates for selected causes of death were calculated for urban and rural populations of each state and correlated with reported percent of residences that had electric service for each state according to urban and rural residence. Urban death rates were much higher than rural rates for most causes, but were poorly correlated with estimated electric service.

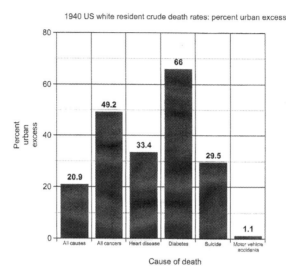

1940 US white resident crude death rates: percent urban excess

FIGURE 13
Urban excess mortality

Rural death rates were strongly and significantly correlated with the level of electric service by state for most of the causes examined. The all-causes mortality rate was similar in urban and rural areas of states with high levels of electrification (above 96 percent), while states with low levels of electrification (below 50 percent) had all-causes death rates at levels twice as high in urban as rural areas.

Nationally, the total cancer rate in 1930 was 58.8 percent higher in urban than in rural areas. This pattern indicates that the twentieth century epidemic of cancers and cardiovascular diseases was not caused by lifestyle variables as is commonly assumed, but rather by EMF exposure, and probably dirty electricity (Milham 2010). The price we have paid for the convenience of electricity in morbidity and mortality since the early 1900s almost defies quantification.

The real surprise of this study was that cardiovascular disease, diabetes, and suicide as well as cancer seem to be strongly related to residential electrification. A community-based epidemiologic study of urban rural differences in coronary heart disease and its risk factors was carried out in the mid-1980s in New Delhi, India, and in a rural area fifty kilometers away (Chadna, et al. 1997). The prevalence of coronary heart disease was three times higher in the urban residents, despite the fact that the rural residents smoked more and had higher total caloric and saturated fat intakes. Most cardiovascular disease risk factors were two to three times more common in the urban residents. Rural electrification projects are still being carried out in parts of the rural area that was studied.

For over eighty years, economists have noted a paradoxical improvement in health indices (declining mortality rates and increasing life expectancy) during economic recessions. Mortality rates increase and life expectancy decreases during economic expansions. The title of a paper (Tapia Granados & Ionides, 2008) "The reversal of the relation between economic growth and health progress: Sweden in the 19th and 20th centuries", sent me into historical electrification and mortality data again.

The expected decline of health indicators with economic recessions and improvement with economic growth in the 19th century Sweden was reversed in the 20th century, giving the counterintuitive pattern of higher mortality and lower life expectancy in economic expansions and improvement of these indices in recessions. The change or "tipping point" occurred at the end of the 19th century or early in the 20th century when electrification was introduced into Sweden.

All 5 of the reversals of annual industrial electric energy use in the US between 1912 and 1970 were accompanied by recessions with lowered GDP, increased unemployment, decreased mortality and increased life expectancy. The mortality improvement between1931 and 1932 by state in the US strongly favored urban (electrified) areas over rural areas. Rural unemployment positively correlated with

residential electrification percentage by state in 1930. The health effects of economic change are mediated by electrical exposure. The improvement of health indices in Nazi occupied Europe in WW II and in Cuba during their recent economic collapse were not due to caloric restriction, but to lowered EMF exposure. See Milham, 2012.

By way of contrast, the Amish in the United States and Canada, who live without electricity, have a current pattern of morbidity and mortality remarkably similar to that of United States rural residents in the early part of the twentieth century. Their type 2 diabetes prevalence rates are about half those of the non-Amish, even though their obesity rates are comparable. Also like the early rural residents, the Amish have lower rates of cancer, cardiovascular disease, and suicide. The life expectancy of the Amish is above seventy years and has been stable since 1890. At the turn of the twentieth century, when almost all United States cities were electrified, urban residents had an average life expectancy below fifty years.

A recent study, "Low cancer incidence rates in Ohio Amish," by an Ohio State University Medical Center group (Westman, et al. 2010), showed that the low cancer incidence in the Amish cannot be explained solely on the basis of their tobacco abstinence or other factors. I'd postulate that another major factor in their low cancer incidence is that many Amish, especially the Old Order Amish (OOA), live without exposure to electricity. Alzheimer's disease has also been reported to have a low prevalence in the Amish. A pediatric group practice, in Jasper, Indiana which cares for more than 800 Amish families, has not diagnosed a single child in this group with attention deficit hyperactivity disorder. Childhood obesity is also virtually nonexistent in this population (Ruff 2005).

I wrote a letter to the journal *Cancer Causes and Control*, which published the Ohio Amish study, suggesting that absence of exposure to electricity might have contributed to the low cancer incidence in the Amish, but the letter was rejected.

If the rest of the US population had the disease incidence

and prevalence of the Old Order Amish, the US medical care and pharmaceutical industries would collapse.

CANCER IN VISTA DEL MONTE ELEMENTARY SCHOOL IN PALM SPRINGS WITH A CELL TOWER ON CAMPUS

In February 2010, I received an E-mail from Kim McClinton, a science teacher at Vista del Monte elementary school in Palm Springs, California. She had heard about the La Quinta study and thought her school had the same problem. The school had a reputation for being a "cancer school" in the school district. Since 2005, there has been a cell phone tower located within a few feet of a classroom wing in the school courtyard.

FIGURE 14
Vista del Monte cell tower

During a visit to the school, I showed Kim how to use the G/S meter, and she produced a color-coded analysis of G/S readings by classroom. The entire school had very high dirty electricity readings. Their dirty electricity levels were higher than those at the La Quinta school. The Vista del Monte G/S readings averaged 1,300 compared to 750 at La Quinta. The cancers (twelve cancers, including six female breast cancers among seventy-five personnel employed at the school since 1990) were over-represented in the wing of the school closest to the cell tower, and the G/S readings were highest in the classrooms closest to the cell tower base. At the same stage of the investigation, La Quinta school had eleven cancers in 137 teachers.

A fourth grade teacher complained that her students were hyperactive and un-teachable. The outlets in her room measured over 5,000 Graham/Stetzer units. On a Friday afternoon after school, I reduced the measured dirty electricity in the wiring from over 5,000 to less than 50 Graham/Stetzer units with five plug-in filters. With no change in either the cell tower radiation or the lighting, the teacher reported an immediate dramatic improvement in student behavior in the following week. They were calmer, paid more attention and were teachable all week except for Wednesday when they spent part of the day in the library. Later, the teacher told me that she could change the behavior of the children by removing and reinserting the filters. The change took between 30 and 45 minutes. This young teacher also became the thirteenth cancer case in this small teachers' cohort.

On January 25, 2011, I presented my findings to the Palm Springs Unified School District Board of Education. I sent the Powerpoint of my presentation in advance. I was surprised to learn at the last minute that the board had hired Leeka Kheifets to contest my findings and had provided her with a copy of my presentation. Of course, I had not been given a copy of her presentation. I offered to filter the school at no cost to the district, guaranteeing an improvement in student test scores and attendance. My offer was refused.

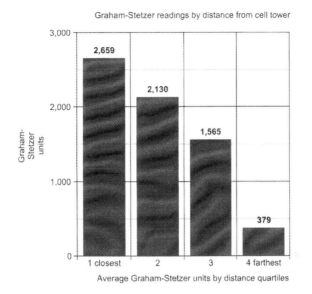

Graham-Stetzer readings by distance from cell tower

FIGURE 15

The thirty-two classroom Graham Stetzer readings were averaged over quartiles ranked by distance from the cell tower

One teacher with poorly controlled type 2 diabetes, in spite of insulin injections and oral hypoglycemic medication, had repeated foot infections and a below-the-knee amputation. He retired in 2009, and his blood glucose readings have been normal since then. Magda Havas has shown that dirty electricity raises blood glucose levels and changes insulin requirement in diabetics. The blood glucose connection could be how dirty electricity increases cardiovascular disease incidence. The major mortality and morbidity in diabetics is due to acceleration of cardiovascular diseases. Magda Havas has also shown that radiation from DECT (Digital Enhanced Cordless Telecommunications) phones can cause an instant change in heart rate and rhythm in some exposed individuals.

Cell tower transmitters, like most modern electrical equipment, operate on direct current. The electrical current brought to the tower is alternating current which needs to be changed to direct current. This is done by a switching power supply or an inverter. These devices interrupt the AC current and are the likely source of the dirty electricity in the wing of the school closest to the tower. That little device you plug into the wall to charge your cell phone is one of these. They are present in all computers, copy machines, and television sets.

This is another serious but unrecognized hazard of cellular telephone technology. People who are concerned about health issues regarding cell towers focus on the RF emissions, but dirty electricity is another unrecognized important exposure.

To illustrate just how far dirty electricity effects can extend, after Dave Stetzer filtered a Midwestern school, a dairy farmer a quarter of a mile away noticed that his cows each gave an average of ten pounds more milk per day beginning the day the school was filtered. The cows were responding to dirty electricity being removed from the ground currents.

URINARY NEUROTRANSMITTERS IN LIBRARIANS

In the summer of 2011, at a book signing at the Olympia Timberland Public Library in Olympia WA, I measured very high levels (>20,000 G/S units) of dirty electricity in the outlets of the room where the book signing took place. The recommended level for no effects is 50 G/S units or less. In the hundreds of dirty electricity measurements made by us over the years, only a very few were this high. Excess mortality in 2,608 female librarians dying in Washington State between 1970 and 2010 was seen for the following causes: cancers of the tongue, breast, ovary and brain, Alzheimers' disease, diseases of the veins and pulmonary embolus. (https://fortress.wa.gov/doh/occmort/). Since the La Quinta school teachers study showed a dose/response between increasing dirty electricity classroom levels and increasing cancer incidence, I felt it important to offer to reduce the dirty electricity exposure of the library employees.

With the permission of the City of Olympia, and the cooperation of the library manager and library staff, on October 10, 2011, Dave Stetzer and I reduced the dirty electricity levels in the library and measured urinary dopamine. and phenylethylamine (PEA) in seven volunteers a few days before and after the clean-up, and every two weeks for the next six months.

The Microsurge meter readings in the same outlet in the manager's office were 11,190 G/S units before the clean-up, 540 G/S units during the clean-up and 39 G/S units at the end of the clean-up which took about two hours. After an initial decrease, over the next six months the average urinary dopamine concentration in the volunteers who provided urine after the clean-up gradually increased to an average of over 225 ug/g creatinine which is well above 170 ug/g creatinine, the high normal level for the lab. Average phenylyethylamine levels also rose gradually to slightly above 70 ug/g creatinine, the high normal level for the lab. After the initial post clean-up urinary sampling, volunteers' homes also had their home dirty electricity levels reduced with filters. These neurotransmitters have been shown to decrease in people who reside near cell towers. (Buchner and Eger, 2011; Eskander et al, 2011). With chronic exposure and stress, neuro-endocrine and immune system dysregulation results in a wide spectrum of human morbidity and mortality. See Milham and Stetzer, 2012.

FUTURE INVESTIGATIONS 10

Here are some areas I believe are worth investigating for possible links with EMFs.

AMYOTROPHIC LATERAL SCLEROSIS (LOU GEHRIG'S DISEASE)

I have a theory regarding the etiology of ALS (amyotrophic lateral sclerosis, popularly called Lou Gehrig's disease, named after the famous baseball player who was afflicted with it.) My theory is that ALS may be caused by exposures to electro-therapeutic devices and elevated EMF/RF environmental exposures.

ALS is a relentlessly progressive neurodegenerative disease, beginning with barely perceptible symptoms like numbness and tingling in the fingers, but leading eventually to total body paralysis. All the while the mind remains intact. Most people with ALS die within five years of diagnosis, with an average survivorship of three years.

The Washington State occupation mortality data shows that ALS has a very high mortality rate in Boeing engineers. On frequent visits to the company twenty years ago, I noticed that most of their engineers worked in large rooms, surrounded closely on four sides by computers. This is a likely high dirty electricity environment, and I would like to see those workspaces accurately measured, by me or someone else.

When I was at the Washington State Department of Health, I helped Boeing set up a mortality surveillance system using their insurance and worker record systems, and helped them resolve disputes with OSHA. My requests to do a study have been ignored.

ALS has been shown to be more common in people who have had serious electric shock episodes, and it is more common in electrical workers in general. For several years now, I have tried to interest others in studying people who have had electro-convulsive shock therapy (ECT), to see if they also develop ALS. Another clue to ALS etiology is clusters of the disease that have been reported in Italian soccer players and in United States football players. These players, and other athletes, have often use transcutaneous electric nerve stimulation devices (TENS), which deliver small electric shocks to the skin to ease pain. It acts as a counter-irritant in the same way as liniment.

Athletes are also often treated with shortwave and microwave diathermy for musculoskeletal injuries. Unlike TENS units, diathermy penetrates deeply into tissue and heats it by using RF, just like a microwave oven. I recently contacted a young woman with ALS (age twenty-seven) who had been a professional basketball player. She reported extensive use of TENS devices beginning in high school and continuing through college and her professional sports career. Her case caught my attention, because ALS is uncommon in people under thirty years of age.

There is some circumstantial evidence that Lou Gehrig, whose name is synonymous with ALS, might have been treated with short wave diathermy. In 1933, Boston Red Sox player Dale Alexander had his baseball career cut short by a serious leg burn and gangrene following a diathermy treatment by a Red Sox trainer named "Doc" Woods, so diathermy was in use by major league baseball teams in the early 1930s.

And a long-time New York Yankees trainer, Earle V. "Doc" Painter, burned Joe DiMaggio's foot with a diathermy treatment during spring training in 1936, DiMaggio's first year with the Yankees. Babe Ruth, Lou Gehrig, and Joe DiMaggio were all reported to have been patients of "Doc" Painter. Plus, Bob Waters, one of the three members of the 1964 San Francisco 49ers professional football team who died of ALS, reported being treated for many hours with diathermy in the team's training room. At least a dozen United States and Canadian professional football players have or have died of ALS.

My hypothesis is that sporadic ALS is caused by exogenous electrical currents induced in, or applied to, the body. Since most cases of ALS have no obvious connection to electrical shock or electrical therapy, most non-familial cases (e.g. cases with no identifiable genetic link) may have had currents induced in their bodies by working or living in environments with strong EMFs. Strong contact currents may also be a contributing exposure. In addition to TENS and diathermy exposure, electro-convulsive shock therapy, electro-surgery, and electric acupuncture also generate electric currents in patients (Milham 2010).

There are many more electrotherapeutics in the development pipelines and nowhere near enough study to conclude these are safe. That's why it is so important to study occupational increases in ALS like those at Boeing for clues.

FALLON, NEVADA CHILDHOOD LEUKEMIA CLUSTER

For a number of years I've been involved in a frustrating attempt to study the Fallon, Nevada childhood leukemia cluster, working with a research group headed by Dr. Joe Weimels of University of California at San Francisco.

Beginning in 1997, seventeen children in the small Nevada oasis town of Fallon (pop. 7,500) have been diagnosed with childhood leukemia. Eleven of the cases were diagnosed between 1997 and 2001, making this the most dramatic childhood leukemia cluster ever reported. This disease is typically caused by fetal exposure during the mother's pregnancy or very early in a child's life, so whatever caused the cluster operated for a very short period of time.

Some clues include: there is a "Top Gun" naval air station close to town, a fuel oil pipeline to the airbase, and a tungsten processing facility in town. The most popular etiologic theories involve agents (tungsten, jet fuel) that have been contaminating the town for a long time.

There are approximately six EMF/RF studies that showed EMF exposures from ground-based antennas, using various frequencies, are associated with childhood leukemia. Yet the Agency for Toxic Substances and Disease Registry (ASTDR) never considered these exposures in its investigation of the cluster. There was a Loran C antenna near the town that I suspect caused the epidemic. The antenna stopped operating in February 2010.

Long after the epidemic peak occurred, a group of local amateur radio operators reported that for a twenty-four hour period around New Year's Day 2006, their transceivers were swamped by the Loran signal. This coincided with a wet weather period, and water makes the earth more conductive to electricity. The Coast Guard, which operated the antennas, admitted to a malfunction. Since it would take a very powerful signal to swamp solid state transceivers twenty-five miles from an antenna, I suspect that the signal travelled through the ground and may have been conducted by the jet fuel pipeline right into town, thereby exposing pregnant women and disabling the amateur radio operators' modern, solid state transceivers. Without actually studying ground current phenomena, we will never know, and no one can do it now that the antenna no longer operates.

Lloyd Morgan, who worked on the La Quinta study with me, also noticed that adult cancers in Churchill County, where Fallon is located, spiked at the same time as the childhood leukemia cluster. There was another wet weather event in the area at about the right time to cause the epidemic. At this late date, twelve years after the first case, we still haven't been able to obtain case address information that was collected in the ASTDR study. .

Fallon has a major dairy industry. If, as I suspect, ground currents were involved in the epidemic, milk production records, over time, might be revealing. Emergency room records and veterinary records should also be studied. This investigation should have happened years ago, not twelve years after the first case.

POWERFUL TRANSMITTERS AND CANCER

For years, the United States Air Force and others have studied the potential health effects of radiofrequency radiation emitted by the Precision Acquisition Vehicle Entry, Phased-Array Warning System known as PAVE PAWS in Barnstable County on Cape Cod, Massachusetts. Recently, there has been concern that the antenna was involved in a cluster of rare Ewing's sarcoma cases near the antenna. The Massachusetts Department of Public Health and the U.S. Air Force have been studying this antenna for more than twenty years.

Though PAVE PAWS has gotten the most attention in the press over the years, there was another powerful antenna in Massachusetts. A (Loran C) antenna on Nantucket Island off the Cape Cod coast was turned off in February 2010. The National Cancer Institute cancer profiles Web site, listed by state, shows that for the fourteen Massachusetts counties from 2001 to 2004, for all races and both sexes, that Barnstable and Nantucket counties have very high cancer incidence rates for total cancers, as well as for certain selected cancers.

Nantucket and Barnstable rank first and second of the fourteen Massachusetts counties for total cancers. Nantucket County ranks first out of the fourteen Massachusetts counties for all cancers, female breast cancer, colon/rectal cancer, and prostate cancer. Being a small population county, data was not available for many uncommon cancers.

Barnstable County ranked second of the fourteen Massachusetts counties for all cancers; first for leukemia and urinary bladder cancer; second for brain cancer, malignant melanoma of the skin, and prostate cancer; and third for female breast cancer.

In addition to the NCI database, the excellent Massachusetts Cancer registry series by city and town allows a closer look for the years 2001 to 2005 at the towns around PAVE PAWS and at Nantucket Island (which is also a town as well as a county). The three towns around PAVE PAWS, Barnstable, Sandwich and Bourne, each have a significant excess of total cancers. Barnstable has a significant excess of malignant melanoma, pancreas and prostate cancer; and Sandwich has a significant excess of malignant melanoma, leukemia, prostate, urinary tract, and bladder cancers.

Nantucket Island has significant elevations of total cancers, malignant melanoma, female breast and colon/rectal cancer, and prostate cancer. I predict that Nantucket Island cancer rates will gradually decline, since the Loran C antenna no longer operates.

Despite the nebulous findings of a National Academy of Sciences report on PAVE PAWS in 2003, there is clearly something going on regarding RF environmental exposures and cancer on Cape Cod. There have also been childhood leukemia clusters near powerful (one million watts) military communications transmitters in Hawaii, Guam, and Scotland, as well as several other clusters near broadcast facilities throughout the world.

POLYCYTHEMIA VERA

In the winter of 2009, I met again with the La Quinta school teachers. Most of them have voted with their feet and left for other schools. The new teachers, unfortunately, have little knowledge of their risk, and they and their students continue to be exposed to the same high levels of dirty electricity as before. The teachers have not been able to secure legal assistance, and are justifiably furious that their union won't help them.

From them, I learned that a second teacher in their group also had a case of polycythemia vera (PCV), in addition to the Burkitt's Lymphoma that killed him. Polycythemia vera is a type of red blood cell leukemia, which can be treated by bloodletting. There were two cases of PCV in that teacher's population, about sixty times what would be expected. Both of the cases taught in rooms that had a maximum reading on the G/S meter (2000 G/S units).

In an online search of PCV, I learned of a PCV cluster in Pennsylvania that was being studied by ASTDR. Four confirmed cases lived on Ben Titus Road in Tamaqua, Pennsylvania. After investigating through Google Earth, I noticed a cogenerating utility plant nearby. These plants burn waste anthracite coal called culm to generate electricity. Culm is deposited in large mounds around coal-producing states, including Pennsylvania. In theory, every electron generated by a utility plant returns to a substation at the plant either through the neutral line of the electric distribution system or through the earth.

I suspect that often, electricity returns through the ground rather than the neutral line, and can therefore get into homes through ground rods and metal plumbing. A clue that high frequency transients might be involved in causing PCV was reported in a 1981 letter to the *New England Journal of Medicine* by Dr. H.L. Freidman. He noted that microwave exposed workers were over-represented among the PCV cases he had examined. The title of his letter was self-explanatory: "Are chronic exposure to microwaves and polycythemia associated?" I sent a G/S meter to a local Tamaqua resident to check the high-frequency voltage transients in the PCV neighborhood. The levels were high. Further study is necessary.

CANCER CLUSTERS

Disease clusters always represent unique opportunities to gain insight into illness and causation.

In the spring of 2009, I learned of a cancer cluster in the Literature Building at San Diego State University (UCSD), which included eight cases of breast cancer since 2000. UCSD epidemiologist, Dr. Cedric Garland, had linked strong electromagnetic fields generated by the building's elevator system to the high rate of cancer in building employees. I contacted three faculty members and the university leadership and offered to investigate the cluster free of charge. Although the faculty was supportive of my participation, the university instead hired Dr. Leeka Kheifets, a long-time consultant for the electric utility industry. Her final report was essentially a literature review of studies already published. Interestingly, Dr. Garland, who did the original study, first learned of her report when I sent him a copy. A new Literature Building committee chairman there has unfortunately shut the door on any further investigation.

Three other cancer clusters in the news recently may well have an EMF connection. A Google Earth flight to the Acreage, Florida shows a high voltage transmission line traversing the area with a utility substation near the school, Cameron, Missouri has a prison north of town with a high voltage electric fence, and Clyde, Ohio has a large Whirlpool appliance assembly plant.

FERTILITY AND ELECTRIFICATION

When conducting the large mortality study using the 1930 and 1940 vital statistics, I noticed that birth rates and fertility were inversely correlated with residential electrification levels; the more electrification, the lower the fertility and birthrates became.

Birth rates and death rates are sometime published in adjacent columns in United States vital statistics books. Third world electrification projects also resulted in lower fertility and birth rates. The people doing this work were economists and sociologists and ascribed the changes to societal and social variables. I think that the changes are due directly to EMF exposures because in one place, the birth rate dropped within a year of electrification. In another place the birth rate was lower in women who didn't practice family planning in an electrified village, compared to women who didn't practice family planning in a village without electricity.

Places in rural China and India are now undergoing electrification. It would be a marvelous opportunity to prospectively study populations being exposed to EMFs for the first time.

Clues to this relationship also exist in animals. There have been several studies that found that cows give less milk with increased EMF exposures, and both cows and pigs have shown fertility problems in such environments.

Dave Stetzer has an interesting fertility anecdote from his area of Wisconsin regarding bank employees. The women bank clerks were having difficulty getting pregnant or were suffering spontaneous abortions. He examined the bank, found high levels of dirty electricity, and reduced them with filters. A year or so later he got an angry call from the bank manager, complaining that many of his workers had left simultaneously on maternity leave. **I think that the boom in fertility clinics may be due to increasing environmental EMF levels.**

TESTICULAR CANCER AND RADAR EXPOSURES

Once in a rare while, all it takes is a letter to decision-makers to make a huge difference for a worker. In 2006, I received a letter from Thomas McCaskell, a retired member of the Royal Canadian Mounted Police, who had developed testicular cancer from holding energized radar guns in his lap. One of my Centers for Disease Control (CDC) trainees, Dr.

Robert Davis, had co-authored a case report showing that police officers who held radar guns in their laps had about seven times the expected risk of testicular cancer. I wrote a letter to the Department of Veterans Affairs in Canada stating my reasons for believing that radar causes testicular cancer and attached a copy of the Davis paper.

About a year later, Mr. McCaskell sent me an E-mail that included a copy of the favorable decision in his case from Veterans Affairs. Most importantly, this sets a precedent for similar cases. Too often, attempts by an outside professional to spontaneously assist in these situations are met with suspicion rather than goodwill.

BRUSHED GENERATORS AND MOTORS

Edison's nine "Jumbo" generators had serious brush arcing problems and commutator wear. This means that from the very beginning of electrification in the US and the rest of the world, dirty electricity was being sent out into the grid. In an attempt to control the arcing, he added metallic mercury to the commutators, but this caused illness in his workers. Brushed generators and motors have the same problem today. At my request, David Stetzer captured wave forms and measured the dirty electricity from three large commercial generators. All had dirty electricity levels above 200 Microsurge meter units and dirty sine waves.

In 2011 the Lancet published a paper (Danaei G. et al, 2011) listing fasting plasma glucose (FPG) and diabetes prevalence in 199 countries and territories around the world. Islands are over-represented in places with high blood glucose and diabetes prevalence. Seven of ten of the places with highest FPG in males are small islands, while only one of the ten places with the lowest FPG are. In 2011, the same group, Global Burden of Metabolic Risk Factors in Chronic Diseases Collaborating Group, also published a similar analysis of body mass index (obesity) (Finucane MM et al, 2011), with nearly identical results. I believe that the world wide epidemics of diabetes and obesity are both due to exposure to dirty electricity on electric utility wiring coming from generator brush arcing, bad wiring connections and from cell tower switching power supplies. Islands without fuel supplies are likely to import diesel oil to fuel generator sets which generate dirty electricity which rides along on the 50 and 60 Hz transmission frequencies. My letter to the Lancet about this was rejected (see my web site).

De-Kun Li (Li DK 2011,2012) has published two important prospective studies showing that magnetic field exposure during pregnancy increases the risk of asthma and obesity in offspring. If the islands of Oceania are cleaned up, it may take a generation to see the effects.

I think I also know the etiology of the excess suicides, post traumatic stress disorders, and a number of other Gulf War illnesses. About 85 percent of the fuel oil imported into Afghanistan and Iraq is used for air conditioning at a cost of $20.2 billion per year (2011 report). The portable diesel-fueled generator sets which power the air conditioners generate a lot of dirty electricity. The wiring also can't be very good, because of the reports of increased accidental electrocution in military personnel in Iraq and Afghanistan. Interestingly, Navy and Air Force personnel don't share the recent suicide increase seen in the Army and Marine Corps.

The highest asthma prevalence rate reported is in the population of Tristan da Cunha, a small Atlantic island with six diesel generator sets for electrical power.

WHAT TO DO

EVER SINCE THE Wertheimer-Leeper 1979 study, the focus of most EMF research has been on the power frequencies of 50- and 60-Hz. These are in the extremely low frequency (ELF) part of the electromagnetic spectrum.

Electricity travels at the speed of light, or about 186,000 miles per second. The wavelength of various frequencies is measured in relationship to the speed of light. For instance, distances in the galaxy are measured in light years, i.e., the distance light travels in a year. Similarly, the wavelength of residential electricity in the United States would be calculated as 186,000 divided by 60-Hz, or 3,200. That means that the distance between successive peaks of a 60-Hz sine wave is more than three thousand miles.

This calculation holds true for all frequencies. The higher the frequency, the shorter the wavelength. For instance, gamma rays have wavelengths of a billionth of an inch. Visible light is also part of the electromagnetic spectrum. The various colors of visible light also differ in wavelength and frequency: red has the longest wavelength and lowest frequency, while violet has the shortest wavelength and highest frequency. Ultraviolet is of course important as our major source of vitamin D.

The energy of EMFs is proportional to frequency. The frequencies above ultraviolet are said to be ionizing, since they have enough energy to knock electrons out of orbit, creating ions. The inadequate existing U.S. safety standards for EMF/RF are based solely on certain frequency's ability to heat tissue the way a microwave oven heats food. The standards acknowledge the importance of frequency in their models because as frequency increases, exposure limits are lowered.

The "smooth" 60-Hz sine wave EMFs seem to have few biological effects in animal exposure studies. If frequencies are pulsed or modulated, they seem to become more bioactive.

When industry apologists say that these fields are too weak to cause biological effects, I point them to any number of electro-therapeutic devices, such as pulsed high-frequency field generators that are used to accelerate the healing of bone fractures. Anything that can stimulate cell division and growth is a potential carcinogen.

HIGH DIRTY ELECTRICITY LEVELS

I carry a G/S meter around and measure outlets whenever I get a chance. I have a modified meter with a maximum reading of 20,000 units, compared to the commercially available meter, which can read 2,000 units. The highest levels I've seen so far are in excess of 20,000 compared to the recommended 50 units. The worst offenders were a lamp store, hospital emergency rooms, and an oncology clinic. A strip mall near a cell tower, a school near a cell tower and a fire station near a cell tower, all have high dirty electricity levels. An electrical outlet in a bicycle repair shop about thirty feet from a cell tower read above 20,000 G/S units. It is ironic that the oncologists, nurses, and staff at an oncology clinic are being exposed eight hours a day where they work to diagnose and treat cancer patients, yet are unaware themselves of their own potential exposures to a controllable carcinogen. Nothing ever came of notifying the hospital facilities people of what I found.

REMEDIES

Most of the devices that generate dirty electricity can be redesigned to eliminate it at very low cost. New buildings should be designed with dirty electricity levels in mind. Existing schools, hospitals, and medical facilities should be surveyed and filtered to reduce dirty electricity levels. Many of the large childhood and adult cancer studies done over the past thirty years should be revisited using the new G/S meter. I predict that their risk ratios will increase. Also a new generation of animal exposure studies should be done using dirty electricity as the new exposure metric.

New case-control studies, especially of malignant melanoma and thyroid cancer, should measure the high frequency transient exposures. These two cancers have among the highest rates of increase in Western populations

In the early days of magnetic field research, the utility companies sponsored magnetic field surveys of residences and offices around the country. Many utility companies will still come to a home or business when asked to take measurements, and we now need to request that they include dirty electricity as part of their surveys.

If dirty electricity can dramatically increase blood sugar levels, it can cause problems in trying to stabilize "brittle" diabetes, the term used to describe poorly controlled blood glucose levels. Childhood asthma has also been linked to dirty electricity exposure in one Wisconsin school. Clinical studies of the effects of dirty electricity exposures in asthma and diabetes are essential. Every school in the country should be looked at and filtered if necessary. With the difficult financial climate, jurisdictions are increasingly leasing school and fire station property for cell tower sites.

Tanning beds should be banned, because they cause malignant melanoma. I don't think it is solely the ultraviolet light that is the carcinogen, but the electrical equipment powering the fluorescent bulbs, too. I have measured very high levels of magnetic fields (above 100 mG) on tanning beds, and high levels of dirty electricity in the outlets of the tanning parlors.

Radial tires can be made with non-magnetic wire. Tire balancing machines can be made to de-magnetize tires as they spin for dynamic balancing.

One easy way to detect higher frequency radiation in your environment is to use a simple portable AM radio, a sensitive detector of dirty electricity. Tune the AM radio to an "off" station, meaning where there is no broadcast occurring, just white noise. Any place in your house or office that generates obvious, increased AM static will indicate high levels of dirty electricity or higher frequencies. Put the radio near a compact fluorescent light and turn the light on and off. Dimmer switches and wireless routers, computers, and copy machines, TV sets and fax machines also generate a lot of dirty electricity. The difference in noise will be obvious. For instance, I bought a halogen lamp which, when plugged in at one end of my house, created static near the outlets at the other end of the house. All AM radio reception, even on the strongest stations, is impossible near the La Quinta school and at a strip mall near a cell tower in Indio, California.

A meter that gives a digital readout of the high frequency voltage transient levels when plugged into electrical outlets is available from Stetzer Electric (*www.stetzerelectric.com/*). There are instructions on how to interpret the numbers, and what ranges are considered safe.

There are also methods to counteract transients that may be coming in over your power line distribution network, or through your ground rods or plumbing, as opposed to what you generate from your own electronic appliances. It is best to consult a licensed electrician before attempting any of this on your own. Sadly, few electricians understand or know how to deal with dirty electricity.

Many companies make and market good meters for measuring various parts of the electromagnetic spectrum. A plug-in capacitive filter is also available from Stetzer Electric to reduce the dirty electricity levels, but removal of the offending devices is a better strategy. Though every location is different, in my experience, removal of twelve compact fluorescent bulbs from a track lighting system allowed me to get below the recommended 50 G/S units, using only about a dozen filters for my entire home in Washington State.

LONG-TERM SOLUTIONS

Unfortunately, funding sources for EMF research have almost completely dried up in the United States, unlike in Europe and other countries. We have also lost a whole generation of young investigators in university, private, and government labs who were forced into other research areas by the lack of money to support this important research. Industry and cell phone money have corrupted the research process, politicians, the media, and the government. After my generation of investigators has moved on, there will be nobody to take our place in the United States.

Ultimately, to get a real handle on the problem, we will have to rethink how we distribute electricity and communicate. Getting rid of wireless forms of communication is a logical first step. I predict that when the latency periods (time between first exposure and diagnosis) for brain tumors have been achieved, we are in for a calamitous epidemic of cell phone-induced brain tumors.

With rare exceptions, such as Lennart Hardell's work in Europe, most of the cell phone brain tumor epidemiology is of such low quality that it doesn't merit publication. With non-participation rates among controls around 50 percent, it is impossible to have any confidence in study results. I suspect that cell phone money has compromised both the investigators and the journals.

Today, homes are being built without wiring for landline telephones. The large telecommunication companies are in a race to force consumers go all wireless. They plan to saturate urban and rural areas with Wi-Fi and Wi-Max systems, and the Federal Communications Commission (FCC) is assisting this with recommendations for private/public (taxpayer) cooperative funding. In addition, broadband Internet connections transmitted over power lines (BPL) have been deployed in certain areas and are spreading quickly. This technology delivers high frequencies to every outlet in the grid.

Many jurisdictions are unfortunately outlawing incandescent light bulbs in favor of the compact fluorescent bulbs (CFLs), which generate RF and high frequency voltage transients in addition to containing mercury. This, too, is a mistake. CFLs will prove more dangerous to the environment and to people than any energy savings they promise. Light emitting diode (LED) bulbs will probably be the best power saving light bulb alternative.

The electric grid should be rebuilt with wiring adequate enough to return currents to the substation without using the ground. The grid was originally built for return currents to travel over wires. It should also be possible to engineer the high-frequency voltage transients out of the electricity that is delivered to our homes and offices.

The trend to "green" energy sources like wind and photovoltaic solar generating facilities may increase exposures to dirty electricity. They use grid intertie inverters and controllers to convert the power they generate into utility-grade electricity that can be sent out on the grid. I expect that commercial fuel cell generating facilities will have the same problem. The devices that do the conversion inject transients onto the grid. **The dirty electricity in a home with a photovoltaic system was 60 G/S units with the inverter turned off and 600 G/S units with it on. Every residential and commercial photovoltaic solar generating facility I've examined generates high levels of dirty electricity.**

Wind farms create two serious health problems; low frequency sound waves (infrasound) caused by the turbine blades and dirty electricity caused by their inverters. Dr. Nina Peirpont has written a book called *Wind Turbine Syndrome,* and Magda Havas and David Colling have a paper describing wind farm dirty electricity (Havas & Colling, 2011).

Similarly, since the electronic equipment of cell towers and terrestrial RF transmitters (AM, FM, TV) operates on direct current, the inverters and switching power supplies that change the AC line voltage to DC, interrupt the current flow and inject dirty electricity into the grid that powers them. I'm sure it is possible to build inverters that won't generate dirty electricity, but until rules and regulations force this, power quality will be sacrificed for profit.

Recent deployment of "smart meters" for residential electricity, water, and gas billing at homes emit RF **and dirty electricity**. The billing information could be sent over existing phone lines, or over fiber-optic lines. Other "smart grid" schemes for self-regulating appliances will also increase residential RF exposure. **Most new appliances have microwave transmitters built into them which communicate constantly with the smart meter.** I believe that the recent increase in the incidence of malignant melanoma and thyroid cancer is due to these fields.

Since most of the electricity that is generated is literally used to turn the wheels of industry with electric motors, it is important that they be made as clean as possible. There is a recent trend toward variable speed motors that generate a lot of dirty electricity. Most new furnace fan motors use variable speed motors to minimize electricity use.

CONCLUSION

The explosive recent increase in radio frequency radiation and high-frequency voltage transient sources, especially in urban areas from cell phones and towers, terrestrial antennas, Wi-Fi and Wi-Max systems, broadband Internet over power lines, and personal electronic equipment, suggests that like the twentieth century EMF epidemic, we may already have a twenty-first century epidemic of morbidity and mortality underway, caused by high frequency electromagnetic fields. The good news is that many of these EMF diseases may be preventable by simple environmental manipulation, if society chooses to pay attention. Unless public outrage intervenes, I'm afraid that our "diseases of civilization" will only get worse. Good science alone is never enough to force sensible public policy. Only citizens can do that.

ʃʃʃʃʃ

APPENDIX

BIOGRAPHICAL INFORMATION

Samuel Milham, MD, attended the Albany NY public schools and earned the BS degree from Union College in Schenectady, New York, the MD degree from Albany Medical College, and an MPH degree from Johns Hopkins University. He is board certified in Public Health, has a Washington State medical license, and has published more than one hundred peer-reviewed articles in his fifty years as a chronic disease epidemiologist. His special interests are: congenital defects, occupational and environmental illness, methods in occupational and environmental studies, and EMF epidemiology.

EDUCATION AND EXPERIENCE

Union College, Schenectady, New York, September 1950–June 1954. B.S. Sigma Xi, Fuller Chemistry Prize

New York State Medical Scholarship

Albany Medical College, September 1954–June 1958, MD. Alpha Omega Alpha

Intern, U.S. Public Health Hospital, Boston, Massachusetts, July 1958–July 1959.

U.S. Public Health Service Residency in Public Health Assigned to Monroe County Health Department, Rochester, New York, July 1959–August 1960.

Johns Hopkins School of Hygiene and Public Health, September 1960–June 1961, MPH.

Senior Resident in Epidemiology, Residency Program, New York State Department of Health, June 1961–1962.

Development Consultant New York State Department of Health, 1963–1967.

Assistant Professor, Department of Pediatrics, Albany Medical College, July 1964–1967.

Diplomate, American Board of Preventive Medicine, June 1966.

Associate Professor, University of Hawaii School of Public Health and Medical School, 1967–1968.

Section Head, Epidemiology, Washington State Department of Health, 1968–1986.

Travel Fellowship, IARC 1971.

Travel Fellowship, International Cancer Research Technology Transfer, 1981.

Washington State Public Health Association Annual Award, 1986.

Chronic Disease Epidemiologist, Washington State Department of Health, 1968–1988.

Clinical Associate Professor, University of Washington School of Public Health, 1968.

Section Head, Chronic Disease Epidemiology Section, Washington State Department of Health, 1988–May 1992.

Adjunct Professor, Mount Sinai School of Medicine, 1989.

Robert Carl Strom Foundation Humanitarian Award, 1990.

Member of Bioelectromagnetics Society, 1984.

Self-employed, June 1992.

Elected to Fellowship, Collegium Ramazzini, October 1994.

Ramazzini Award, 1997.

RESOURCES

For information about all aspects of EMF, *Microwave News* has no peer. *Microwave News* also lists meters for measuring EMFs (http://www.microwavenews.com/).

Magda Havas's work is online at http://www.magdahavas.com/.

The Bioiniative Report reviews the rationale for EMF and RF exposure standards: www.bioinitiative.org/report/index.htm

Catherine Kleiber : http://www.electricalpollution.com

Janet Newton, EMR Policy Institute: www.emrpolicy.org

"Dirty Electricity" meters, filters, and plug strips are available from Stetzer Electric (http://www.stetzerelectric.com/).

I like the Tri-Field Meter from Alpha Labs (http://www.trifield.com/).

Less EMF Inc (www.lessemf.com/) has about anything else you might need for measuring or shielding EMFs.

Radio Shack sells a $15 portable AM/FM radio, model no. 12-586, which is a very sensitive detector of dirty electricity over a wide range of frequencies in the AM band.

Following are some recommended books for the lay reader. Many are available from Less EMF:

Becker, Robert O. and G. Selden. *The Body Electric: Electromagnetism and the Foundation of Life.* New York, NY. William Morrow and Company, 1985.

Brodeur, Paul. *The Great Powerline Coverup. How the Utilities and the Government Are Trying to Hide the Cancer Hazards Posed by Electromagnetic Fields.* New York, NY. Little, Brown and Company,1993.

Havas, Magda and Camilla Reese. *Public Health SOS: The Shadow Side of The Wireless Revolution.* Wide Angle Health, 2009.

Levitt, B. Blake. *Electromagnetic Fields: A Consumer's Guide to the Issues and How to Protect Ourselves.* New York, NY: Harcourt Brace, 1996. (Also available in an iUniverse edition, 2007.)

Sugarman, Ellen. *Warning, the Electricity Around You May Be Hazardous For Your Health.* New York, NY. Simon and Schuster, 1992.

Abstracts of most of my publications are available online at the National Library of Medicine Web site: Google "Pubmed" and key "Milham S" in the blank space.

Electronic copies of my recent publications and my complete CV are available from me at: smilham@dc.rr.com **or at my website: www.sammilham. com.**

REFERENCES

Armstrong, B., G. Theriault, Guenel, J. Deadman, M. Goldberg, and P. Heroux .1994. "Association between exposure to pulsed electromagnetic fields and cancer in electric utility workers in Quebec and France." *American Journal of Epidemiology* 140, no. 9 (Nov.1, 1994):805–20.

Ausubel, J.H. and C. Marchetti. "Elektron: Electrical Systems in Retrospect and Prospect." *Daedalus* 125, no 3 (1996):139–169.

Buchner K., Eger H. (2011). Changes of Clinically Important Neurotransmitters under the Influence of Modulated RF Fields—A Long-term Study under Real-life Conditions. *Umwelt-Medizin-Gesellschaft* 24(1): 44-57. Original in German.

Chadna, S.L., N. Gopinath, and S. Sheckhawat. "Urban-rural differences in the prevalence of coronary heart disease and its risk factors." *Bulletin of the World Health Organization* 75, no. 1 (1997): 31–38.

Court-Brown, W.M. and R. Doll. "Leukemia in childhood and young adult life: Trends in mortality in relation to aetiology." *British Medical Journal* 26, (1961): 981–988.

Danaei G, Finucane MM, Lu Y et al. 2011. National, regional, and global trends in fasting plasma glucose and diabetes prevalence since 1980: systematic analysis of health examination surveys and epidemiologic

studies with 370 countries and 2.7 million participants. *Lancet*; 378: 31-40.

Davis, R.L. and S. Milham. "Altered immune status in aluminum reduction plant workers." *American Journal of Industrial Medicine* 18, (1990):79–85.

Eskander E.F., Estefan S.F., Abd-Rabou A.A. (2011). How does long term exposure to base stations and mobile phones affect human hormone profiles. *Clinical Biochemistry*, In press.

Finucane MM, Stevens GA, Cowan MJ et al. (2011) National, regional, and global trends in body-mass index since 1980: systematic analysis of health examination surveys and epidemiological studies with 960 country-years and 9·1 million participants. *Lancet.* 377: 557-67.

Gittlesohn, A. and S. Milham. "The declining incidence of central nervous system anomalies in New York State." *British Journal of Preventative & Social Medicine* 16, no. 3 (1962): 153--58.

Gittlesohn, A. and S. Milham. "Observations on twinning in New York State." *British Journal of Preventative & Social Medicine* 19, no. 1 (1965): 8–17.

Havas, M. and Colling, D.(2011) Wind turbines make waves: Why some residents near wind turbines become ill. *Bulletin of Science Technology and Society* XX (X) 1-13.

Li, DK, Chen H, Odouli R. Maternal exposure to magnetic fields during pregnancy in relation to the risk of asthma in offspring. *Arch. Pediatr Adolesc. Med.* 2011; 165, 945-950.

Li, DK, Ferber, J, Odouli R et al. Prospective Study of *In-utero* Exposure to Magnetic Fields and the Risk of Childhood Obesity. *Scientific Reports* 2012 July doi:10.1038/srep00540.

Milham, S. "Increased incidence of anencephalus and spina bifida in siblings of affected cases." *Science* 138, no. 3540 (1962): 593–94.

Milham, S. "Leukemia clusters." *Lancet* 2 (1963):1122.

Milham, S. "Pituitary gonadotrophin and dizygotic twinning." *Lancet* 2 (1964): 566.

Milham, S. "Leukemia in husbands and wives." *Science* 148, no. 3666 (1965): 98–100.

Milham, S. and J. Hesser. "Hodgkin's disease in woodworkers." *Lancet* 2, no. 7507 (1967):136–37.

Milham, S. and W. Elledge. "Maternal Methimazole and congenital defects in children." *Teratology* 5, (1972): 125.

Milham, S. "A study of the mortality experience of the AFL-CIO United Brotherhood of Carpenters and Joiners of America." 1969–1970. DHEW Publication No. (NIOSH), 74–129.

Milham, S. "Occupational mortality in Washington State." 1950–1971. DHEW Publication No. (NIOSH) 76–175 A.

Milham, S. "Studies of morbidity near a copper smelter. Env. Health Perspectives. Vol. 19 (1977): 131–132.

Milham, S. "Cancer in aluminum reduction plant workers." *Journal of Occupational and Environmental Medicine* 7, (1979):475–480.

Milham, S. "Mortality from leukemia in workers exposed to electrical and magnetic fields." *New England Journal of Medicine* 307, no. 4 (1982): 249.

Milham, S. "Increased mortality in amateur radio operators due to lymphatic and hematopoeitic malignancies." *American Journal of Epidemiology* 127, no. 1 (1988):50–54.

Milham, S. "Mortality by license class in amateur radio operators." *American Journal of Epidemiology* 128, no. 5 (1988): 1175–1176.

Milham, S. "Increased cancer incidence in office workers exposed too strong magnetic fields." *American Journal of Industrial Medicine* 30 (1996): 702–704.

Milham, S., J. Hatfield, and R. Tell. "Magnetic fields from steel-belted radial tires: implications for epidemiologic studies." *Bioelectromagnetics* 20 (1999): 440–445.

Milham, S. and E.M. Ossiander. "Historical evidence that residential electrification caused the emergence of the childhood leukemia peak." *Medical Hypotheses* 56, no. 3 (2001): 290–295.

Milham, S., and E.M. Ossiander. "Electric typewriter exposure and increased female breast cancer mortality in typists." *Medical Hypotheses* 62, no. 2 (2007): 450–451.

Milham, S., and E.M. Ossiander. "Low proportion of male births and low birth weight of sons of flour mill worker fathers." *American Journal of Industrial Medicine* 51, no. 2 (Feb 2008):157–158.

Milham, S. and L.L. Morgan. "A new electromagnetic exposure metric: high frequency voltage transients associated with increased cancer incidence in teachers in a California school." *American Journal of Industrial Medicine* 51 (2008): 579–586.

Milham, S. "Most cancer in firefighters is due to radio-frequency radiation exposure not inhaled carcinogens." *Medical Hypotheses* 73 (2009): 788-789.

Milham, S. "Historical evidence that electrification caused the 20th century epidemic of diseases of civilization." *Medical Hypotheses* 74, no. 2 (2010): 337–345.

Milham, S. "Amyotrophic lateral sclerosis (Lou Gehrig's disease) is caused by electric currents applied to or induced in the body: it is an iatrogenic disease of athletes caused by use of electrotherapy devices." *Medical Hypotheses* 74 (2010): 1086-7.

Milham, S. Evidence that Dirty Electricity is Causing the World-Wide Epidemics of Obesity and Diabetes. *Electromagnetic Biology and Medicine*. 2012 in Press.

Milham, S. Hypothesis: The reversal of the relation between economic growth and health progress in Sweden in the 19th and 20th centuries was caused by electrification. *Electromagnetic Biology and Medicine*. 2012 in Press.

Milham, S. and Stetzer, D. "Dirty Electricity, Chronic Stress, Neurotransmitters and Disease". *Electromagnetic Biology and Medicine*. 2012 in Press.

Reynolds, P., E.P. Elkin, M.E. Layefsky, and J.M. Lee. "Cancer in California school employees." *American Journal of Industrial Medicine* 36 (1999): 271–278.

Ruff, M.E. "Attention deficit disorder and stimulant use: an epidemic of modernity." *Clin Pediatr* (Phila). 2005 Sep;44(7):557-63.

Szmiegelski, S. "Cancer morbidity in subjects occupationally exposed to high frequency (radio frequency and microwave) electromagnetic radiation." *Science of the Total Environment* 180, no. 1 (Feb 2, 1996): 9–17.

Tapia Granados, J.A.,Ionides, E.L., (2008).The reversal of the relation between economic growth and health progress: Sweden in the 19th and 20th centuries. *Journal of Health Economics* 27,544-563.

Westman, J.A., A.K. Ferketich, R.M. Kauffman, et al. "Low cancer incidence rates in Ohio Amish."*Cancer Causes and Control* 1 (2010): 69-75.